HANDBOOK OF
SEAFLOOR SONAR IMAGERY

WILEY–PRAXIS SERIES IN REMOTE SENSING
Series Editor: Professor David Sloggett, M.Sc., Ph.D.
Visiting Professor, School of Humanities, Kings College London
Co-Director, Dundee Centre for Coastal Zone Research, UK
Deputy Chairman, Anite Systems, UK
Editor, *EARSeL Newsletter*

This series aims to bring together some of the world's leading researchers working in the forefront of the analysis and application of remotely sensed data and the infrastructure required to utilize the data on an operational basis. A key theme of the series is monitoring the environment and the development of sustainable practices for its exploitation.

The series makes an important contribution to existing literature encompassing areas such as: theoretical research; data analysis; the infrastructure required to exploit the data; and the application of data derived from satellites, aircraft and *in situ* observations. The series specifically emphasizes research into the interaction of elements of the global ecosystem publishing high-quality material at the forefront of existing knowledge. It also provides unique insights into examples where remotely sensed data is combined with Geographic Information Systems and high-fidelity models of the physical, chemical and biological processes at the heart of our environment to provide operational applications of the remotely sensed data.

Aimed at a wide readership, the books will appeal to professional researchers working in the field of remote sensing, potential users of the information and data derived from the application of remote sensing techniques, and postgraduate and undergraduate students working in the field.

HANDBOOK OF SEAFLOOR SONAR IMAGERY
Philippe Blondel and Bramley J. Murton, Southampton Oceanography Centre, Southampton, UK

EARTHWATCH: The Climate from Space
John E. Harries, Professor of Earth Observation, Imperial College, London, UK

GLOBAL CHANGE AND REMOTE SENSING
Kirill Ya. Kondratyev, Scientific Research Centre for Ecological Safety, Russian Academy of Sciences, St Petersburg, Russia; A. A. Buznikov, Electrotechnical University of St Petersburg, Russia; O.M. Pokrovsky, Main Geophysical Observatory, St Petersburg, Russia

HIGH LATITUDE CLIMATE AND REMOTE SENSING
Kirill Ya. Kondratyev, Scientific Research Centre for Ecological Safety, Russian Academy of Sciences, St Petersburg, Russia; O.M. Johannessen, Nansen Environment and Remote Sensing Centre, Bergen, Norway; V.V. Melentyev, Nansen International Environmental Remote Sensing Centre, St Petersburg, Russia

REMOTE SENSING AND GEOGRAPHIC INFORMATION SYSTEMS: Geological Mapping, Mineral Exploration and Mining
Christopher A. Legg, United Kingdom Overseas Development Administration, Forest and Land Use Mapping Project, Forest Department, Colombo, Sri Lanka

SATELLITE OCEANOGRAPHY: An Introduction for Oceanographers and Remote-sensing Scientists
Ian S. Robinson, Department of Oceanography, University of Southampton, UK

GEOGRAPHIC INFORMATION FROM SPACE: Processes and Applications of Geocoded Satellite Images
Jonathan Williams, Consultant, Space Division Logica plc, Leatherhead, UK

Forthcoming titles
MULTI-DIMENSIONAL GLOBAL CHANGE
Kirill Ya Kondratyev, Councillor for the Russian Academy of Sciences, Research Centre for Ecological Safety, St Petersburg, Russia

PASSIVE MICROWAVE REMOTE SENSING OF OCEANS
Victor Yu Raizer and Igor V. Cherny, Centre for Program Studies, Russian Academy of Sciences, Moscow, Russia

REMOTE SENSING OF TROPICAL REGIONS
Eugene A. Sharkov, Head of Remote Sensing Laboratory, Space Research Institute, Russian Academy of Sciences, Moscow, Russia

HANDBOOK OF SEAFLOOR SONAR IMAGERY

Philippe Blondel and Bramley J. Murton
Southampton Oceanography Centre
Southampton, UK

JOHN WILEY & SONS

Chichester • New York • Weinheim • Brisbane • Singapore • Toronto

PRAXIS

Published in association with
PRAXIS PUBLISHING
Chichester

Copyright © 1997 Praxis Publishing Ltd
The White House,
Eastergate, Chichester,
West Sussex, PO20 6UR, England

Published in 1997 by
John Wiley & Sons Ltd
in association with Praxis Publishing Ltd

Wiley Editorial Offices

John Wiley & Sons Ltd, Baffins Lane,
Chichester, West Sussex PO19 1UD, England

John Wiley & Sons, Inc., 605 Third Avenue,
New York, NY 10158-0012, USA

VCH Verlagsgesellschaft mvH, Pappelallee 3
D-69469 Weinheim, Germany

Jacaranda Wiley Ltd, G.P.O. Box 859, Brisbane,
Queensland 4001, Australia

John Wiley & Sons (Asia) Pte Ltd, 2 Clementi Loop #02-01,
Jin Xing Distripark, Singapore 129801

John Wiley & Sons (Canada) Ltd, 22 Worcester Road,
Rexdale, Ontario M9W 1L1, Canada

Library of Congress Cataloging-in-Publication Data
Blondel, Philippe.
   Handbook of Seafloor Sonar Imagery / Philippe Blondel and Bramley J. Murton.
   p. cm. -- (Wiley-Praxis series in remote sensing)
   Includes bibliographical references and index.
   ISBN 0-471-96217-1
   1. Ocean bottom--Remote sensing. 2. Sonar. I. murton B. J. . II Title. III Series.
GC87.B47. 1996
551.46'084--dc21   96-47603
A catalogue record for this book is available from the British Library   CIP

ISBN 0-471-96217-1

Printed and bound in Great Britain by Hartnolls Ltd, Bodmin

List of Contributors

Chapter 1
Introduction

Philippe Blondel and Bramley J. Murton

Chapter 2
Sonar Data Acquisition

Philippe Blondel

Chapter 3
Sonar Data Processing

Philippe Blondel

Chapter 4
Deep-Ocean Trenches
and Collision Margins

Bramley J. Murton

Chapter 5
Mid-Ocean Ridge Environments

Bramley J. Murton

Chapter 6
Abyssal Plains and Basins

Philippe Blondel

Chapter 7
Continental Margins

Philippe Blondel

Chapter 8
Coastal Environments

Doris Milkert and Veit Hühnerbach

Chapter 9
Image Anomalies and
Sonar System Artefacts

Philippe Blondel

Chapter 10   Philippe Blondel
Computer-Assisted Interpretation

Chapter 11   Philippe Blondel and Bramley J. Murton
Conclusion

Table of Contents

List of Contributors   v
Preface   xiii
Acknowledgements   xv
List of Tables   xvi
List of Illustrations   xvii

1. INTRODUCTION   1

2. SONAR DATA ACQUISITION   5
2.1 Acoustic Remote Sensing   5
2.2 Sidescan Sonar Imagery   8
2.3 Performance of Sidescan Sonar Systems   10
2.4 Technical Specifications of Commonly Used Sonars   11
2.5 Navigation and Attitude   16
2.5.1 Ship Navigation   16
2.5.2 Towfish Navigation   18
2.5.3 Attitude Information   20
2.6 Summary   20
2.7 Further Reading   21

3. SONAR DATA PROCESSING   23
3.1 Introduction   23
3.2 Pre-Processing   24
3.2.1 Data Formats   24
3.2.2 Navigation Data Processing   24
3.2.3 Attitude Data Processing   26
3.3 Radiometric Corrections   27
3.3.1 Requantisation   27
3.3.2 Across-Track Corrections   28
3.3.3 Along-Track Corrections   29

ı

3.4 Geometric Corrections   30
3.4.1 Slant-Range Correction   30
3.4.2 Anamorphosis   31
3.5 Backscattering Models   32
3.6 Map Production   34
3.6.1 Mosaicking - Stencilling   34
3.6.2 Interpolation - Rubbersheeting   35
3.7 Post-Processing   36
3.7.1 Image Statistics   36
3.7.2 Histogram Manipulations   38
3.7.3 Speckle Removal   39
3.7.4 Sea Surface Reflection Removal   41
3.8 Examples of Operational Sidescan Sonar Processing 41
3.8.1 Near Real-Time Processing of TOBI Data   41
3.8.2 Processing of High-Resolution DSL-120 Data   42
3.9 Summary   43
3.10 Further Reading   45

4. DEEP-OCEAN TRENCHES AND COLLISION MARGINS   **47**
4.1 Plate Tectonics and the Seafloor Environments   48
4.2 Geological Background   49
4.3 The Reasoning behind the Interpretation   51
4.4 Deformation Fronts and Accretionary Prisms   53
4.5 Trench-Fill Structures and Processes   59
4.6 Mud Volcanoes and Serpentinite Seamounts   65
4.7 Summary   69
4.8 Further Reading   70

5. MID-OCEAN RIDGE ENVIRONMENTS   **73**
5.1 Geological Background   73
5.2 Shape Derivation From Backscatter   75
5.3 Volcanic Features   78
5.3.1 Point-Source Volcanoes   79
5.3.2 Composite Volcanoes   82
5.3.3 Central Volcanoes   83
5.3.4 Flat-Topped Volcanoes   87
5.3.5 Clustered Volcanoes   90
5.3.6 Hummocky Volcanic Ridges   96
5.3.7 Axial Volcanic Ridges   102
5.4 Tectonic Features   106
5.4.1 General Tectonic Trends at Mid-Ocean Ridges   106
5.4.2 Single Faults and Fissures   107
5.4.3 Fault Scarps   109
5.4.4 Tectonic Features at Transform and Non-Transform Offsets   113
5.5 Hydrothermal Features   116
5.5.1 Hydrothermal Mounds   117
5.5.2 Hydrothermal Structures   119

5.6 Regional Imagery   120
5.6.1 West Lau Spreading Centre   120
5.6.2 Central Indian Ocean Triple Junction   123
5.7 Further Reading   126

6. ABYSSAL PLAINS AND BASINS   **129**
6.1 Geological Background   130
6.2 The Use of Subsurface Information   131
6.3 The Filling of Abyssal Plains   133
6.4 Structures of the Abyssal Plains   137
6.5 Relict Structures   144
6.6 Example of Regional Imagery   146
6.7 Summary   151
6.8 Further Reading   152

7. CONTINENTAL MARGINS   **153**
7.1 Geological Background   154
7.2 Sedimentary Structures   155
7.2.1 Sediment Deposition and Erosion   155
7.2.2 Sediment Transport - Submarine Canyons   157
7.2.3 Mass-Wasting, Slides and Flows   163
7.2.4 Sediment Redistribution   167
7.3 Tectonic Structures   170
7.4 Examples of Regional Imagery   171
7.4.1 The Blake Escarpment, North Atlantic   171
7.4.2 The East Arequipa Basin, South Pacific   174
7.5 Volcanic Structures   177
7.6 Biological Activity   179
7.7 Structures From The Epicontinental Seas   181
7.7.1 Mud Volcanism   181
7.7.2 Brine Accumulation Structures   186
7.7.3 Pockmarks and Seepages   188
7.8 Summary   190
7.9 Further Reading   191

8. COASTAL ENVIRONMENTS   **193**
8.1 Geological Background   194
8.2 Sedimentary Cover   194
8.3 Rock Outcrops   197
8.4 Sedimentary Features Created by Waves and Currents   200
8.4.1 Ripples   201
8.4.2 Megaripples   202
8.4.3 Longitudinal Bedforms   204
8.4.4 Obstacle Marks   205
8.5 Glacial Features   206

8.5.1 Iceberg Ploughmarks   206
8.5.2 Slump Structures in the Arctic Environment   206
8.6 Pockmarks   207
8.7 Biological Activity   210
8.7.1 Corals   210
8.7.2 Seafloor Vegetation   212
8.7.3 Schools of Fish   212
8.8 Anthropogenic Structures   213
8.8.1 Anchor Tracks   213
8.8.2 Trawl Marks   214
8.8.3 Wrecks   216
8.8.4 Dump Sites   219
8.8.5 Other Anthropogenic Features   220
8.9 Summary   221
8.10 Further Reading   221

9. IMAGE ANOMALIES AND SONAR SYSTEM ARTEFACTS 223
9.1 Introduction   223
9.2 Water Column Artefacts   224
9.3 Radiometric Artefacts   225
9.4 Geometric Artefacts   227
9.4.1 Variations in Survey Speed   227
9.4.2 Variations in the Platform's Altitude - Heave   228
9.4.3 Unprocessed Roll   230
9.4.4 Unprocessed Pitch   231
9.4.5 Unprocessed Yaw   231
9.5 Processing and Output Artefacts   233
9.5.1 Beam Spreading   233
9.5.2 Slant-Range Corrections - Layover   234
9.5.3 Processing Artefacts   236
9.5.4 Output Artefacts   237
9.6 Interpretation Artefacts   238
9.6.1 Subsurface Reflections   238
9.6.2 Interference Effects   238
9.6.3 Multiple Reflections   240
9.6.4 Unexpected Features   242
9.7 Conclusion   245
9.8 Further Reading   246

10. COMPUTER-ASSISTED INTERPRETATION   247
10.1 Introduction   247
10.2 Image Enhancement   248
10.2.1 Image Representation   248
10.2.2 Image Statistics   251
10.2.3 Image Enhancement Techniques   252
10.2.3.1 Histogram Manipulation   252
10.2.3.2 Image Smoothing   253

10.2.3.3 Image Sharpening   255
10.3 Contour-Oriented Analysis   256
10.3.1 Spot Structures   256
10.3.2 Contour Detection   257
10.3.3 Contour Analysis   260
10.4 Texture-Oriented Analysis   262
10.4.1 Texture Definition   262
10.4.2 Structural Methods   264
10.4.3 Statistical Methods   264
10.5 Fusion of Multi-Source Information   265
10.5.1 Bathymetric Information   265
10.5.2 Other Information - Ground-Truthing   267
10.6 Geographic Information Systems   267
10.6.1 GIS Definition   267
10.6.2 GIS Composition   268
10.6.3 GIS Operation   269
10.6.4 Examples of GIS Applications   269
10.6.4.1 Crustal-Scale Fissuring in the East Pacific   269
10.6.4.2 Distribution of Hydrothermal Deposits in the North Atlantic   269
10.7 Image Classification - Image Compression   271
10.7.1 Image Classification   271
10.7.2 Image Compression   273
10.8 Artificial Intelligence - Expert Systems   274
10.9 Summary   275
10.10 Further Reading   276

11 CONCLUSION   277

General Bibliography   279
Glossary   305
Index   307

The colour plate section appears between pages 106 and 107

Preface

The challenges of marine exploration and exploitation have highlighted the need for sidescan sonar. Because it is less attenuated by water than other remote sensing techniques, sonar is the only tool capable of accurately mapping large areas of the seafloor. Accordingly, it is now used everywhere around the world, in waters often uncharted. Despite being used for several decades in all marine environments, until now there has been no comprehensive book on the use of sidescan sonar.

Our idea for the present "Handbook of Seafloor Sonar Imagery" was inspired by the need to fill the gap in the literature. This book is divided into three main sections: the acquisition of sonar imagery, examples from the different environments (deepest to shallowest) and techniques of advanced interpretation. We have tried to make it widely accessible by pitching it at a scientific graduate level. No book on sidescan sonar imagery has been written since the 1970s. However, the past decades have seen great advances in sidescan sonar and its processing. We felt, therefore, that it was timely to provide users with an up-to-date reference work.

The research described in this publication was carried by the authors while at the Institute of Oceanographic Sciences (Wormley, UK) and later at the Southampton Oceanographic Centre under contracts with the Natural Environment Research Council. Reference herein to any specific commercial product, process, or service by trade name, trademark, manufacturer, or otherwise, does not constitute or imply its endorsement by the British Government, the Natural Environment Research Council, or the Southampton Oceanography Centre.

Acknowledgments

This book is the first comprehensive review of sidescan sonar imagery of the seafloor. It was based on our own research and activities, but has benefited from the influence of many people. In particular, we would like to thank our Project Manager, Lindsay Parson, for his understanding and discreet but unfailing assistance. The pioneering work of many of our colleagues around the world, and in particular at SOC and its predecessor the Institute of Oceanographic Sciences, were influential in paving the way for the writing of this book. Among many others too numerous to cite here, we would like to thank Mike Somers, Doug Masson, Neil Kenyon, Tim LeBas, Roger Searle, Neil Mitchell, R.A. Jablonski, F.J. Hollender, and Rachel Cave.

We were very pleased to include in this book a chapter dedicated to coastal environments and written by Doris Milkert and Veit Hühnerbach, from the University of Kiel. They are excellent friends and colleagues, and have been very helpful at the different stages of the book's preparation.

We would like to thank the following for their thorough reviews: Dave Coller and Martin Critchley (ERA-MAPTEC, Dublin), Eric Pouliquen (SACLANTCEN, La Spezia), F. Werner (University of Kiel), A. Kuijpers and R. Köster (Germany), Doug Masson (SOC) and Mike Shardlow (PRAXIS).

Finally, we would like to thank all the scientists around the world who provided us with high-quality sidescan sonar images. Some of them did not hesitate to open their "treasure chests" for us to pick and print the images we thought were the most appropriate, including some previously unpublished data: Doug Masson (SOC), Graham Westbrook (University of Birmingham), Rick Hagen (Alfred-Wegener Institut), V. Purnachandra Rao and M. Veerayya (NIO, India) and Valerie Paskevich (USGS). Many others made particular images available, including Joe Cann, Sandy Shor, Susan Humphris, Marty Kleinrock, Ken Stewart, Keith Pickering, Bill Schwab, Patty Fryer, Tim LeBas, B. Bader, H.G. Schröder, K. Schwarzer, P. Schäfer, K.W. Tietze and A. Wehrmann.

List of Tables

Chapter 2

2.1	Characteristics of the most commonly encountered deep-sea sonars	12
2.2	Characteristics of the most commonly encountered shallow-water sonars	13
2.2	Characteristics of the most commonly encountered shallow-water sonars (continued)	14
2.3	Characteristics of multibeam systems providing sidescan sonar imagery	15

Chapter 10

10.1	Texture analysis techniques found in the literature	263
10.2	Additional source of interpretation which can help the interpretation	266
10.3	Basic problems that can be investigated with GIS	269

List of Illustrations

Chapter 1
1.1   ETOPO5 world topography   2
1.2   Topographic profile across South America and the Atlantic   3

Chapter 2
2.1   Frequencies in use by acoustic systems   6
2.2   Definition of some parameters   7
2.3   Interaction of the acoustic pulse with the seafloor   8
2.4   Formation of an image by acquisition of cross-track profiles   9
2.5   Definition of the footprint   11
2.6   Satellite navigation systems   17
2.7   Layback of the towfish   18
2.8   Short baseline system   19
2.9   Long baseline system   19

Chapter 3
3.1   Respective positions of the survey vessel and the towfish   25
3.2   Along-track variations of attitude during an actual survey   26
3.3   Attenuation of backscatter signals with distance   28
3.4   Examples of along- and across-track striping   29
3.5   Slant-range distortion   30
3.6   Anamorphosis   31
3.7   Factors influencing the backscattering   32
3.8   Example of a backscattering model   33
3.9   Mosaicking and stencilling   34
3.10   Overlaps and incomplete coverage during ship's turns   35

3.11   The aspect of an image can be described by its statistics   37
3.12   Example of histogram equalisation   38
3.13   Example of speckle   39
3.14   Multiple reflections on the sea surface   40
3.15   Data processing chain   44

Chapter 4
4.1   Location of deep-ocean trenches and collision margins   47
4.2   Section through the Earth showing mantle, core and crust   49
4.3   Perspective view of an oceanic subduction zone   50
4.4   GLORIA imagery of the Columbian Trench   55
4.5   Interpretation of the GLORIA imagery   56
4.6   GLORIA imagery of the floor of the Columbian Trench   57
4.7   Interpretation of the GLORIA imagery   57
4.8   IZANAGI imagery of the Nankai Trough   58
4.9   Interpretation of the IZANAGI imagery   59
4.10   GLORIA imagery of the Columbian Trench   60
4.11   Interpretation of the GLORIA imagery   61
4.12   GLORIA imagery of a collision margin   62
4.13   Interpretation of the GLORIA imagery   63
4.14   GLORIA imagery of the Puerto Rico Trench draped over   64
         bathymetry
4.15   GLORIA imagery of mud volcanoes   66
4.16   Interpretation of the GLORIA imagery   67
4.17   SeaMARC II imagery of a serpentinite seamount   68
4.18   Interpretation of the SeaMARC II imagery   69

Chapter 5
5.1   Location of mid-oceanic ridges around the world   73
5.2   DSL-120 imagery of the TAG hydrothermal mound   76
5.3   Histograms of pixel values across-track   76
5.4   TOBI imagery of the western wall of the Mid-Atlantic Ridge   77
5.5   Histograms of pixel values across-track   78
5.6   TOBI imagery and interpretation of a point-source volcano   79
5.7   SeaMARC II imagery and interpretation of a point-source volcano 80
5.8   TOBI imagery and interpretation of a composite volcano   82
5.9   TOBI imagery of the W-seamount on the Mid-Atlantic Ridge   84
5.10   Interpretation of the TOBI imagery   85
5.11   Skyline profile of the W-seamount   86
5.12   TOBI imagery of the W-seamount draped over bathymetry   86
5.13   TOBI imagery of a flat-topped seamount   88
5.14   Distortion of elevated objects by lay-over   88
5.15   Interpretation of the TOBI imagery   89
5.16   SeaMARC II imagery of clustered volcanoes   91
5.17   Interpretation of the SeaMARC II imagery   91
5.18   TOBI imagery of the same clustered volcanoes   92
5.19   Interpretation of the TOBI imagery   92
5.20   TOBI imagery of clustered volcanoes   94
5.21   Interpretation of the TOBI imagery   95

5.22	TOBI imagery of a hummocky volcanic ridge	97
5.23	Interpretation of the TOBI imagery	98
5.24	Shape profiles derived from the TOBI imagery	98
5.25	Skyline profiles	99
5.26	SeaMARC I imagery of hummocky volcanic ridges	100
5.27	Interpretation of the SeaMARC I imagery	101
5.28	TOBI imagery of an axial volcanic ridge	102
5.29	Interpretation of the TOBI imagery	103
5.30	GLORIA imagery of hummocky ridges	104
5.31	TOBI imagery of the same hummocky ridges	105
5.32	SeaMARC II imagery of simple faults	107
5.33	TOBI imagery of the same simple faults	108
5.34	TOBI imagery of fault scarps	109
5.35	Interpretation of the TOBI imagery	110
5.36	TOBI imagery and interpretation of major axial valley wall fault splays	111
5.37	SeaMARC I imagery of fault splays	113
5.38	TOBI imagery of a non-transform discontinuity	114
5.39	Interpretation of the TOBI imagery	115
5.40	Photographs of hydrothermal "black smokers"	116
5.41	AMS-120 and TOBI imagery of the TAG hydrothermal mound	117
5.42	Interpretation of the AMS-120 imagery	118
5.43	Unscaled shape profile of the imagery	118
5.44	Schematic cross-section of the TAG mound	119
5.45	DSL-120 imagery and interpretation of hydrothermal vents	120
5.46	GLORIA imagery of the northern Lau Basin	121
5.47	Interpretation of the GLORIA imagery	122
5.48	GLORIA imagery of the Central Indian Ocean Triple Junction	124
5.49	Interpretation of the GLORIA imagery	125
5.50	GLORIA imagery draped over multibeam bathymetry	126

Chapter 6

6.1	Abyssal plains and basins throughout the world	129
6.2	Physiography of the abyssal plains	130
6.3	Examples of typical 3.5-kHz profiles	132
6.4	GLORIA imagery of the Mississippi Fan	134
6.5	GLORIA imagery and interpretation of the Gulf of Mexico	135
6.6	TOBI imagery of blocks in the El Golfo debris avalanche	136
6.7	Close-up views of TOBI imagery	137
6.8	TOBI imagery of fresh turbidites	138
6.9	TOBI imagery of a debris slide and an older turbidity channel	139
6.10	TOBI imagery of a slump	139
6.11	TOBI imagery of flow-banding sub-parallel to the sonar	140
6.12	TOBI imagery of flow-banding oriented obliquely to the sonar	141
6.13	TOBI imagery of pressure ridges	141
6.14	TOBI imagery of longitudinal shear structures	142
6.15	TOBI imagery of erosional structures	143
6.16	TOBI imagery of an old (100 Ma) seamount	144
6.17	TOBI imagery of a relict block	145

6.18   Martian analogue of a relict block deflecting the flows   146
6.19   GLORIA imagery of debris avalanches off Hawaii   147
6.20   Close-up view of the different geological units   149
6.21   Interpretation of the GLORIA imagery   150

Chapter 7
7.1   Continental margins and marginal seas throughout the world   153
7.2   Physiography of the continental margins   154
7.3   GLORIA imagery and interpretation of a current-scoured outcrop 156
7.4   GLORIA imagery and interpretation of sediment transport   158
7.5   GLORIA imagery of sedimentary channels   159
7.6(a) GLORIA imagery of Veatch Canyon   160
7.6(b) Interpretation of the GLORIA imagery   161
7.7   Close-up view of a part of Veatch Canyon   161
7.8   SeaMARC II imagery and interpretation of cut-off meanders   162
7.9   GLORIA imagery and interpretation of debris slides   164
7.10   SeaMARC I imagery of a teardrop slide   165
7.11   Stylised profiles of turbiditic deposits   166
7.12   SeaMARC II imagery of terraces   167
7.13   SeaMARC II imagery of arcuate sediment waves   168
7.14   GLORIA imagery of linear sediment waves   169
7.15   TOBI imagery of slumps   170
7.16   MAK-1 record of a fault scarp   171
7.17(a) GLORIA imagery of the Blake Escarpment   172
7.17(b) Interpretation of the GLORIA imagery   173
7.18   SeaMARC II imagery of the central East Arequipa Basin   175
7.19   SeaMARC II imagery of the southern East Arequipa Basin   176
7.20   GLORIA imagery of Bear Seamount   177
7.21   GLORIA imagery of Knaus Knoll   178
7.22   Sidescan imagery of *Halimeda* bioherms   179
7.23   Sidescan imagery of massive algal biohermal structures   180
7.24   Linear algal ridges   181
7.25   MAK-1 imagery of mud vents and mud flows   182
7.26   MAK-1 imagery of a mud volcano   183
7.27   EG&G 990S imagery of a mud volcano   184
7.28   Interpretation of the EG&G 990S imagery   185
7.29   MAK-1 imagery of brine pools   186
7.30   TOBI imagery of L'Atalante Basin   187
7.31   MAK-1 imagery of pockmarks   189

Chapter 8
8.1   Location of coastal environments throughout the world   193
8.2   Coarse lag sediment over eroded glacial till   195
8.3   Mud and sand deposits   196
8.4   Featureless muddy sediments with otter-trawl tracks   197
8.5   Outcropping stratified bedrock   198
8.6   Outcropping metamorphic bedrock   199
8.7   Outcropping Eemian clay and sand ripples   200
8.8   Wave ripples in medium-grained sand   201

8.9	Ripples organised along channels	202
8.10	Sandy megaripples	203
8.11	Megaripples composed of fine carbonate sand	203
8.12	Sand ribbons	204
8.13	Current-induced scours and moats around obstacles on the seafloor	205
8.14	Iceberg ploughmarks in a glacial fjord	206
8.15	Slump structures in the Arctic environment	207
8.16	Elongated pockmark in the Baltic Sea	208
8.17	Funnel-shaped pockmarks in a lake	209
8.18	Funnel-shaped pockmarks in a lake	210
8.19	Coral mounds on Sula Ridge (Norway)	211
8.20	Old and new coral colonies on Sula Ridge (Norway)	211
8.21	Seagrass on a sandy seafloor	212
8.22	Schools of fish in the water column	213
8.23	Anchor tracks on a muddy harbour floor	214
8.24	Trawl marks on a muddy seafloor	215
8.25	Trawl marks on a sandy seafloor	216
8.26	Shipwreck of a wooden coal transport vessel	217
8.27	Partially buried shipwreck	217
8.28	Wooden wreck near a shingle beach	218
8.29	Dump site on a muddy background	219
8.30	Stone circles, possibly of Neolithic origin	220

Chapter 9

9.1	The "Face on Mars" and the "Face on the Seafloor"	223
9.2	Near-nadir artefacts	226
9.3	Rapid attenuation of backscatter when the platform goes up	227
9.4	Examples of correct and incorrect speed corrections	228
9.5	Variations in the platform's altitude - Heave	229
9.6	Example of unprocessed heave	230
9.7	Unprocessed roll	231
9.8	Unprocessed pitch	231
9.9	Unprocessed raw	232
9.10	Beam spreading	233
9.11	Example of beam spreading	234
9.12	Illustration of the layover effect	235
9.13	Examples of inappropriate TVG and AVG corrections	236
9.14	Example of across-track artefact	237
9.15	GLORIA record of interference fringes	239
9.16	TOBI imagery of interference fringes	240
9.17	Formation of multiples	241
9.18	SeaMARC II imagery of multiple reflections on the seafloor	242
9.19	Example of an unexpected feature	243
9.20	The "Face on Mars" and the "Face on the Seafloor"	244
9.21	Example of an unexpected feature	245

Chapter 10

10.1	Examples of sampling size reduction	249

10.2   Examples of reduction in the number of grey levels   250
10.3   Skewness and kurtosis   252
10.4   Histogram manipulation with "control points"   253
10.5   Examples of smoothing   254
10.6   Examples of sharpening   255
10.7   TOBI imagery of spot structures   257
10.8   DSL-120 imagery of linear structures   258
10.9   Filtering of a TOBI image with Kirsch gradients   259
10.10   Adaptive filtering of a TOBI image   259
10.11   Tracking methods   261
10.12   Examples of specific searching patterns   262
10.13   Examples of different textures   263
10.14   Simple features may be discernible on the basis of grey levels 264
          alone
10.15   Stylised comparison of vector and raster representations   268
10.16   Schematic representation of a neural network   274

Colour plate section (between pages 106 and 107)

1   DSL-120 imagery and bathymetry from the TAG mound
2   TOBI imagery draped over bathymetry from the W-seamount
3   GLORIA imagery of the Indian Ridge Triple Junction draped over bathymetry
4   Perspective view of GLORIA imagery of the Indian Ridge Triple Junction
5a   Perspective view of TOBI imagery and bathymetry of the Mid-Atlantic Ridge
5b   Plan view of TOBI imagery of the Mid-Atlantic Ridge
6   First-order classification of TOBI imagery from a non-transform discontinuity
7   Textural analysis and classification of TOBI imagery from a discontinuity

1

Introduction

Knowledge of the Earth and its environment are proving increasingly crucial. Scientific, economic, politic and social decisions all depend at some time or another on this knowledge and we like to think that we know everything about our planet. One may be justified in doing so today by looking back to those maps with white unexplored regions that were still prevalent until the beginning of the 20th century. Yet as we move into the 21st century, in many respects we know more about the solid surface of other planets than about our own Earth. The recent Magellan probe orbited Venus for several years and gave us a precise map of the whole planet's surface with resolutions ranging from 75 to 200 metres. But we are far from knowing the landscapes that lie only a few kilometres from our shores. The unknown and uncharted can be found just a few hundred kilometres from our major cities !

This is because more than two-thirds of the Earth's surface are covered by oceans and thus not easily accessible to direct observation. It is only in the last 10 to 20 years that technological advances have allowed us to discover and map the Earth's seafloor, via acoustic remote sensing and submersibles. Dives are limited in time, geographic coverage and depth. Only a handful of submersibles in the world are capable of diving lower than 3,000 m (i.e. 46% of the planet's surface). Sound waves are the only means of surveying large regions of the seafloor with a reasonable accuracy. Being easily transmitted through the water column, their reflection off the seafloor allows us to image its shape and texture.

Acoustic waves are the basis of sonars (SOund Navigation And Ranging). The images and maps produced by these sensors are not always easy to interpret, because of their nature and because of the complex processes at play during the reflection of the acoustic waves on the seafloor. No comprehensive book explaining the different steps of sonar imagery interpretation has been published since the early days of the 1970s (Belderson et al., in 1972) (see General Bibliography, Chapter 8). Yet, since then, huge advances have been made in acoustic remote sensing technology, its applications, and the understanding of natural phenomena. We felt, therefore, that it was timely to provide both end-users of sonars and specialists with a reference text they would be able to use frequently.

Who should be concerned with this book? Everyone, in fact. We have endeavoured to write it for a scientific graduate level to accommodate the wide range of potential readers who expressed their interest in such a book. Students and teachers can use this book during university and engineering classes concerned with acoustic remote sensing and surveying of the seafloor as well as in marine geology. Specialists of marine data acquisition can consult it when confronted with unknown features, to refresh their knowledge about certain processes, or to keep abreast of the latest developments in sonar imagery. Marine scientists will find here the latest information about sonar data processing and interpretation, from the recently discovered deep-sea hydrothermal vents to the applications of artificial intelligence and expert systems. As oil exploration moves to deeper waters, environmental impact studies are increasing and sonar systems are getting more widespread, every potential user should find something for their field of application in this book. Even readers who are interested in the lesser-known domains of our planet will enjoy this book. The images in it represent some of the latest views from "inner space" and have an aesthetic quality that is as rare and exciting as any from our solar system.

Figure 1.1. ETOPO5 world topography (5'-resolution).

What is this book concerned with? Everything, or nearly everything on the seafloor. This book aims at covering all the stages from data acquisition to interpretation and decision-making. Sonar data acquisition (Chapter 2) is treated in detail from the transmission of the acoustic wave to its interaction with the seafloor and the recording of the backscattered wave. The next chapter (Chapter 3) explains how the successive waves are processed to form images and maps, and presents real-life examples of detailed processing performed at sea. The next chapters show the applications, i.e. representative images from the different domains accessible with sonar, along with the reasoning behind their interpretation. These chapters have been divided into the different physiographic regions. Figure 1.1 shows the global topography of the Earth with a 5'-resolution (around 10 km at the equator) as compiled by the National Geophysical Data Center (United States). These measurements are represented with different grey levels. The darkest levels correspond to the greatest seafloor depths and they progressively brighten as the depth increases. Even at this resolution, very distinct regions can be identified: deep flat basins, curvilinear shallower features like the mid-oceanic ridges, and very deep trenches (Figure 1.2). In a progression from the deepest to the shallowest, one can find deep-ocean trenches (Chapter 4), mid-ocean ridges (Chapter 5), abyssal plains and basins (Chapter 6), continental slopes and shelves (Chapter 7) and coastal environments (Chapter 8). A specific section has been devoted to image anomalies (Chapter 9), showing the stages where artefacts can appear, how they can be avoided during the processing, and how to recognise and interpret them when they occur. It draws both on the most recent theoretical studies on these subjects, and on real-world examples in a variety of applications. The last section presents the state of the art in computer-assisted interpretation (Chapter 10): image processing techniques, artificial intelligence, and the emergence of expert systems. These chapters are supplemented with a bibliography, covering in greater detail the different subjects and directing the reader to sources of more focused information, and a glossary explaining the technical terms and specialists' jargon.

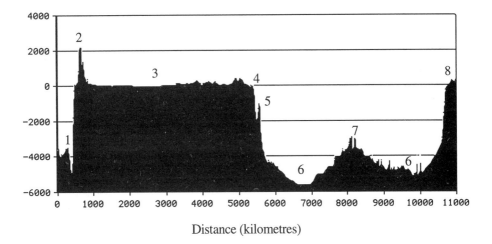

Distance (kilometres)

Figure 1.2. West-East Profile showing the different geological provinces: deep-sea trench [1], the Andes [2] and the South-American continent [3], the Brazilian continental shelf [4] and the continental slope [5], the abyssal plains [6], the Mid-Atlantic Ridge [7] and the symmetric rise leading to Africa [8].

This book does not intend to present all types of sonar images available, but only those that we consider the most representative. Despite the enormous amount of sonar imagery collected by our co-workers from the Southampton Oceanography Centre, we have tried as much as possible to present imagery from other types of sonar as well. It is certain that some acoustic systems will not be referred to (hopefully the less frequently used). And it is certain that a few rare geological processes will not be mentioned either ... as said earlier, this book is the first comprehensive "Handbook of Seafloor Sonar Imagery" since the 1970s. All suggestions for improvement will be welcome, and we hope the present edition will be found useful in all fields of applications, academic and industrial alike.

2

Sonar Data Acquisition

2.1 ACOUSTIC REMOTE SENSING

Acoustic remote sensing is the only means to map and study the surface morphology of the seafloor at all depths. Conventional remote sensing methods (optical or radar) fail because of the high absorption of electromagnetic waves in the water. Current advances in physics and electronics have allowed the development of all sorts of acoustic transducers with frequencies ranging between a few hertz and several megahertz (Figure 2.1). Acoustic waves are easily generated, not much absorbed in the water column, and their reflection off the seafloor gives details about local morphology. In this chapter, we will endeavour to present the basics of acoustic remote sensing, the acquisition of sidescan sonar imagery, the performance and technical specifications of the sonar systems most commonly used, and we will also present the state-of-the-art techniques for positioning the sonars.

Sonar systems can be roughly divided into three categories: echo-sounders, sidescan sonars, and multibeam sonars. Echo-sounders transmit one single beam, oriented vertically. Sidescan sonars generally transmit two beams, one on each side (although the first models were only transmitting on one side). And multibeam sonars transmit several tens of beams on each side. These systems may acquire bathymetry, or acoustic imagery, or both. The distinction is somewhat more blended now as the most recent generation of multibeam systems also acquire sidescan imagery. These different systems can be hull-mounted, shallow-towed, deep-towed, or autonomous. Echo-sounders and multibeam systems are mostly hull-mounted, and sidescan sonars towed behind the ship.

Sound waves propagate in water with a velocity of approximately 1500 m/s. Local variations occur, from place to place or from time to time, due to variations in water

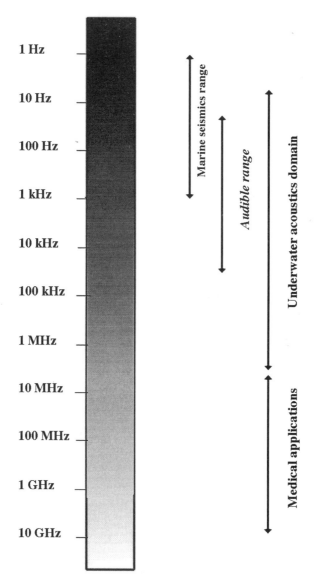

Figure 2.1. Frequencies in use by different acoustic systems

temperature, salinity, and hydrostatic pressure. Numerous empirical formulae are available in the specialised literature (e.g. Brekhovskikh and Lysanov, 1991; journals such as the *Journal of the Acoustical Society of America*). Updated calculations are also available in the numerical databases from oceanographic institutions (e.g. the Levitus database from Woods Hole, Massachusetts, U.S.A.). These variations define layers in the water column. The incoming sound waves are refracted as they cross these interfaces, and their trajectories can be accurately assimilated to slightly bending rays. They are also attenuated with the distance, depending with the frequency and the local concentrations of magnesium sulphate. The propagation of acoustic waves is also affected by the presence of physical heterogeneities (marine organisms, such as plankton in the Deep Scattering Layer, 100-300 m below sea-level, or gas bubbles near the surface), reverberation from the sea surface, and sea noise (thermal noise with the radiation from warm layers to cold layers; surface waves; animal and human-produced noises such as passing whales or ships). These effects have varying contributions depending on the frequency used and the region surveyed.

The acoustic beams transmitted by the sonar interact with the seafloor (Figure 2.2) and most of their energy is reflected specularly. The distance travelled from the acoustic transducer to the target of the seafloor is called the "slant-range". It is not to be confused with the horizontal distance ("ground range") between the sonar's nadir and the target. The angle between the incoming wave and the seafloor is called the "angle of incidence". The grazing angle is 90° minus the incidence angle, i.e. the angle between the incoming wave and the local normal to the seafloor. The angle between the sonar's nadir and the incident beam is called the "elevation angle" or the "look angle". The sonar platform (ship of other vehicle) is travelling along a trajectory or track, sometimes called "line-of-sight" in other domains of remote sensing. The total distance ensonified across-track is the "swath width".

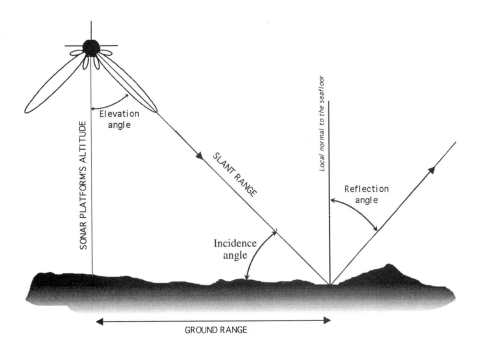

Figure 2.2. Definitions of some parameters.

2.2 SIDESCAN SONAR IMAGERY

Most of the energy arriving on to the seafloor is scattered forward in the specular direction. A small portion is lost in the ground, and a small portion (several orders of magnitude smaller than the incident wave) is scattered back to the sonar, amplified and recorded (Figure 2.3). The timeshift between the transmission and the reception is directly proportional to the distance between the sensor and its target (slant-range). The frequency shift indicates the speed of the target relative to the sensor (Doppler effect), but is not useful in geological applications. Phase shifts are used to deduce the arrival angle of the beam. Some systems also extract bathymetry from the phase difference between neighbouring transducers. And the amplitude is related to the amount of backscattering on the seafloor. The backscattering is affected, in decreasing order of importance, by :

- the geometry of the sensor-target system (angle of incidence of each beam, local slope, etc.)
- the physical characteristics of the surface (micro-scale roughness, ...)
- the intrinsic nature of the surface (composition, density, relative importance of volume vs. surface diffusion for the selected frequency)

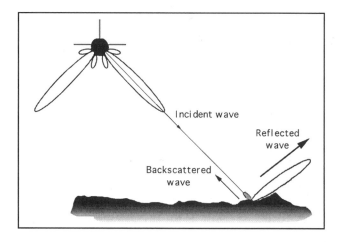

Figure 2.3. Interaction of the acoustic pulse with the seafloor.

For each transmission cycle, or ping, the received signal is recorded over a relatively long time-window, such that the energy returned from across a broad swath of seafloor is stored sequentially. This cross-track scanning produces individual profiles like the one in the Figure 2.4 (top). The successive along-track profiles are accumulated and, put together, form maps of the seafloor's backscatter (Figure 2.4, bottom).

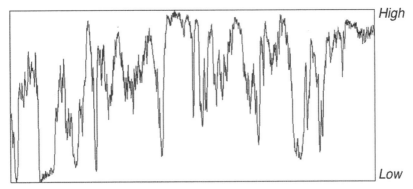

Individual cross-track profile of backscatter

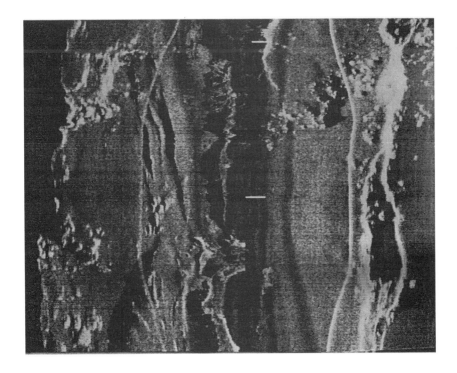

Figure 2.4. This TOBI image of the Reykjanes Ridge is formed by the acquisition of successive individual cross-track profiles of backscatter like the one above.

2.3 PERFORMANCE OF SIDESCAN SONAR SYSTEMS

An important factor in the performance of a sidescan sonar system is its resolution. Unfortunately, several definitions co-exist: the resolution may be the area ensonified by each beam, the spacing between successive measurements, the pixel size in images or the minimum scale at which one one object on the seafloor can be detected or two distinct objects separated.

The three-dimensional distribution of acoustic energy creates ellipses on the ensonified areas of the seafloor. The area covered by each ellipse is called a "footprint" (Figure 2.5). The across-track footprint of the sonar is determined by the length of the transmitted pulse projected onto the seafloor and is given by the equation :

$$\Delta x = \frac{c}{2} \times \frac{L}{\cos\theta} = \frac{c}{2B}$$

where L is the pulse length (in seconds), c the sound velocity and θ the grazing angle. The parameter B often appears in the literature and is the bandwidth of the signal. The angular variation of Δx means that the across-track footprint gets progressively poorer towards the nadir. The along-track footprint Δy is either the width of the horizontal beam on the ground or the distance travelled by the transducer during the reception interval, whichever is smallest. At close ranges, the bottom is often undersampled and some features may be missed. This is one of the reasons why the sonar tracks are always hard to interpret[1]. The along-track footprint increases linearly with the slant range. The aspect ratio increases with the distance, in extreme cases leading to the impression that all far-range bottom targets are sub-parallel to the track.

The across-track spacing between data points corresponds to the swath width divided by the number of points actually recorded. It is related to the sampling digitisation rate. The along-track spacing corresponds to the time between each successive ping. It is directly related to the survey speed and the transducer's capabilities. Areas on the seafloor will be oversampled if the spacing is smaller than the footprint, and undersampled otherwise. Most sidescan systems rely on the oversampling scheme.

The incorporation of the successive across-track profiles into images requires the computation of square picture elements ("pixels"). The pixel size can be anything. If greater than the distance between measurements, the different backscatter values are averaged (smoothing and loss of small-scale details). If the pixel size is chosen to be smaller, interpolation is needed. Whatever the interpolation algorithm chosen, this is more likely to produce errors by creating new data or distorting existing features.

---

[1] The other reason is that, because of technical constraints, the beams present sidelobes that may interfere with each other (cross-talk).

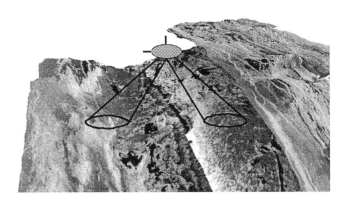

Figure 2.5. The portion of the seafloor ensonified by the sonar is called the footprint.

From these images, single features may be detected on the basis of one pixel only (and therefore the resolving power is the same as the pixel size). This is not totally true, as the pixel could be influenced by one small portion of the area ensonified or some electronic noise, and confident manual interpretation usually requires several pixels. The separation of two distinct features will likewise require several pixels.

One important point to keep in mind is that the performance of sidescan sonars should not be reduced to the resolution of the system. Many other aspects intervene, such as transmitting power and frequency, transducer length and design, pulse repetition rate, survey speed, height above the seafloor, etc. All these aspects should be taken into consideration when planning a survey or comparing datasets acquired on the same area.

2.4 TECHNICAL SPECIFICATIONS OF COMMONLY USED SONARS

To allow the comparison of existing sidescan sonar systems, we have compiled a short list, showing the technical specifications of several types of sonar. For each instrument, the operating institutions have been noted, as well as a broad generic description of the sensor (i.e. sidescan or multibeam, deep-towed or hull-mounted, etc.). The wavelength λ has been computed from the classical formula $\lambda = c / v$ where c is the sound propagation velocity (1500 m.s^{-1}) and v is the transmission frequency.

	GLORIA Mk II	GLORIA - B	TOBI	SeaMARC II / HMR-1	SAR 190 kHz	DSL-120	Jason 200 kHz
Operator(s)	SOC, USGS, ...	SOC	SOC	Univ. of Hawaii	IFREMER	WHOI	WHOI
Type	Shallow-tow	Shallow-tow	Deep-tow	Shallow-tow	Deep-tow	Deep-tow	Deep-tow
Depth range	200 - 11,000 m	200 - 11,000 m	< 10,500 m	100 - 11,000 m	< 6,200 m	< 6,100 m	< 6,000 m
Total swath width	up to 60 km (typically 45)	45 km (imagery) 4.5 x depth (bathymetry)	6 km	up to 10 km (typically 10 km)	up to 1.5 km	0.1 to 1 km (typically 1 km)	0.3 km
Typical navigation error	100 - 1000 m satellite	100 - 1000 m satellite	50-500 m long-baseline	100 - 1000 m satellite	5 - 10 m long-baseline	5 - 10 m long-baseline	0.1 - 10 m short- or long-baseline
Typical daily coverage	20,000 km²	20,000 km² (imagery) 11,500 km² (bathymetry)	470 km²	10,000 km²	10 km²	90 km²	10 km²
Frequency (wavelength)	6.3 - 6.7 kHz 23.8 - 22.4 cm	6.25 - 6.75 kHz 24 - 22.2 cm	30-32 kHz 5 - 4.7 cm	11-12 kHz 13.6 - 14.9 cm	170 - 190 kHz 0.9 - 0.8 cm	120 kHz 1.25 cm	200 kHz 0.75 cm
Footprint size (along- x across-track)	175 x 45 to 657 x 45 m	125 x 45 to 1000 x 45 m	8 x 3.5 m to 43 x 2.1 m	120 x 10 m to 197 x 2 m	0.7 x 0.8 m to 3 x 0.4 m	3.3 x 0.33 m to 13.7 x 0.15 m	0.5 x 0.29 m to 2.4 x 0.15 m
Output data	Imagery	Imagery Bathymetry	Imagery	Imagery Bathymetry	Imagery	Imagery Bathymetry	Imagery Bathymetry
Typical size of daily data	< 1 Gbyte	~ 5.3 Mbyte	528 Mbyte	< 1 Gbyte	< 1 Gbyte	> 1 Gbyte	~ 100 Mbyte
Ancillary data	Heading	Heading + Roll, pitch, yaw	Heading, depth, speed + Roll, pitch, yaw	Heading + Roll, pitch, yaw	Heading, depth, speed	Heading, depth + Roll, pitch, yaw	Heading, depth + Roll, pitch, yaw

Table 2.1. Characteristics of the most commonly encountered deep-sea sonars.

	[TAMU]²	EG&G Deep-Tow	EG&G model 272T	Klein 590/595	Klein 520	Simrad MS-992
Operator(s)	Texas A&M Univ.	Government and commercial surveys	Government and commercial surveys	Klein Assoc. Massachusetts, U. Kiel, ...	U. Kiel, Klein Assoc., etc.	Commercial, military ...
Type	Shallow-tow	Deep-tow	Shallow-tow	Shallow-tow	Shallow-tow	Deep-tow
Depth range	< 500 m	< 600 m		< 1,000 m	< 300 m	< 1,000 m
Total swath width	100 m - 30 km	< 1 km	N/A	< 600 m (100 kHz) < 400 m (500 kHz)	25 m - 600 m	10 - 800 m
Typical navigation error	< 100 m satellite + near-shore reckoning	< 100 m satellite + near-shore reckoning	< 100 m satellite + near-shore reckoning	< 100 m satellite + near-shore reckoning	< 100 m satellite + near-shore reckoning	< 100 m satellite + near-shore reckoning
Typical daily coverage	< 13,000 km²	Dependent on survey speed	Dependent on survey speed	Dependent on survey speed	Dependent on survey speed	Dependent on survey speed
Frequency (wavelength)	11/12 kHz 72 kHz	59 kHz	105 kHz	100 kHz 500 kHz	500 kHz	120 kHz 330 kHz
Footprint size (along- x across-track)	Not available	1/400 of the range	N/A	N/A	N/A	N/A
Output data	Imagery Bathymetry	Imagery	Imagery	Imagery	Imagery	Imagery
Typical size of daily data	< 1 Gbyte					
Ancillary data	Depth + Roll, pitch, yaw	Speed, depth, temperature				N/A

Table 2.2. Characteristics of some commonly encountered shallow-water sonars (*continues*).

	OKEAN	MAK-1	EG&G990S	Widescan 60	AMS-36 /120S1	Sys09 (SSI Int'l.)
Operator(s)	CIS States	CIS States	Commercial applications, GPI Kiel	Commercial applications, SOC	Acoustic Marine Systems, Inc.	Commercial applications
Type	Shallow-tow	Deep-tow	Deep-tow	Shallow-tow	Deep-tow	Shallow-tow
Depth range				< 300 m	< 6,000 m	60 - 10,000 m
Total swath width	2 x 8,000 m	2 x 1,000 m or 2 x 250 m	2 x 400 m	37.5 m to 400 m	≤ 1,000 m	< 20 km
Typical navigation error	100 - 1000 m satellite	SBL net	< 100 m satellite + near-shore reckoning	< 100 m satellite + near-shore reckoning	100 - 1000 m satellite	100 - 1000 m satellite
Typical daily coverage	600 km²	Dependent on survey speed	Dependent on survey speed	Dependent on survey speed	Dependent on survey speed	Dependent on survey speed
Frequency (wavelength)	9.5 kHz	30 kHz or 100 kHz		100 kHz/ 325 kHz	33.3 / 36 kHz	9/10 kHz
Footprint size (along- x across-track)	100 x 5 m	35 x 0.5 m	N/A	~ 0.2 m	< 1 m	0.2 x 0.2 m to 10 x 10 m
Output data	Imagery	Imagery	Imagery	Imagery	Imagery	Imagery Bathymetry
Typical size of daily data	≤ 350 Mbyte	≤ 1 Gbyte	Paper record	« 1 Gbyte	≤ 1 Gbyte	> 1 Gbyte
Ancillary data	N/A	N/A	Pressure, heading, speed, temperature	Altitude	Altitude, roll, pitch, yaw	N/A

Table 2.2 (continued). Characteristics of some commonly encountered shallow-water sonars.

	SeaBeam 2000	Hydrosweep	Simrad EM-12	Simrad EM-100	Simrad EM-1000
Operator(s)	SIO, JMSA, KORDI	UW[1], LDEO[2], Germany, India, others ...	SIMRAD A/S, RVS, IFREMER ...	SIMRAD A/S, Canadian Hydrographic Service ...	SIMRAD A/S, Canadian Hydrographic Service ...
Type	Hull-mounted	Hull-mounted	Hull-mounted	Hull-mounted	Hull-mounted
Depth range	200 - 11,000 m	10 - 10,000 m	10-11,000 m	10-700 m	3-1,000 m
Total swath width	90° - 120°	2 x water depth (less if depth > 7 km)	150° (7.3 x water depth) 90°-120° for EM-12S	150° (7.3 x water depth)	150° (7.3 x water depth)
Typical navigation error	100 - 1,000 m (satellite positioning)	100 -1,000 m (satellite positioning)	100 - 1,000 m (satellite positioning)	100 - 1,000 m (satellite positioning)	100 - 1,000 m (satellite positioning)
Typical daily coverage	2,500 km²	3,700 km²	4,000 km²	Dependent on survey speed	Dependent on survey speed
Frequency (wavelength)	12 kHz (12.5 cm)	15.5 kHz 9.7 cm)	13 kHz (11.5 cm)	95 kHz (1.6 cm)	95 kHz (1.6 cm)
Footprint size (along- x across-track)	3.5% x 3.5% water depth	134 x 134 m	170 x 170 m	170 x 170 m	170 x 170 m
Output data	Bathymetry + backscattering amplitude	Bathymetry + backscattering amplitude	Bathymetry + backscattering amplitude	Bathymetry + backscattering amplitude	Bathymetry + backscattering amplitude
Typical size of daily data	< 1 Gbyte	~ 1 Gbyte	Not available	Not available	Not available
Ancillary data	Heading + Roll, pitch, yaw	Heading + Roll, pitch	Heading + Roll, pitch, yaw	Heading + Roll, pitch, yaw	Heading + Roll, pitch, yaw

[1] University of Washington, Seattle, Wa., USA.
[2] Lamont-Doherty Earth Observatory, Palisades, N.Y., USA.

Table 2.3. Some multibeam systems also provide sidescan imagery and pseudo-sidescan imagery (backscatter values associated to each beam).

All figures concerning swath width and daily coverage may change according to the settings chosen by the operator, the depth, and the speed of the ship. Values shown here indicate the ranges, or typical values. The typical amount of daily data is computed from these values, or derived from technical reports. It is also indicative only of a typical setting. Finally, ancillary data include (apart from ship's navigation) the attitude parameters available for advanced pre-processing. References are scattered, and come, either from personal communications from users of these systems, or from recent review articles, such as Kleinrock (1992). This list only includes sonar systems which have been shown to have applications in marine geology. It is in no way intending to rate their relative performances (see last section). And it must be emphasized that it is far from being exhaustive, the near-daily manufacturing of new or updated sonars precluding this. These three tables only show some sonar characteristics in order to give the reader an idea of their possibilities.

2.5 NAVIGATION AND ATTITUDE

To precisely locate the different sonar echoes on the seabed, one needs precise location information: where is the sonar located and which area of the seafloor is it ensonifying? This information is given by the combination of position information (latitude, longitude, altitude above a reference datum), heading, speed, and attitude information (heave, roll, pitch, yaw). The next subsections will detail the techniques currently used for measuring the position of the survey vessel (i.e. at the sea surface), the position of the sonar platform if not attached to the survey vessel (i.e. below the sea surface), and their relative variations in attitude.

2.5.1 Ship Navigation

Ship positioning techniques have drastically changed in ease and accuracy since the "GPS revolution" in the early 1990s. The earlier navigation methods were of course the sextant and, for near-shore surveys, triangulation with artificial or natural landmarks on-shore. Navigation was only known in sparse points, with a limited accuracy depending on the user's experience. Operations occurring further off-shore or requiring a better precision make use of radionavigation. The first type of radionavigation aids use the travel time between the ship and a shore station (examples are radar, shoran, and the range-range mode of Loran-C). More complex systems measure the difference in travel time (or phase) of signals coming from several stations (examples are Raydist, Lorac, Decca, Toran, Loran and Omega), or combine distance and direction measurements through microwaves. Systems like shoran are effective up to 80 km from the shore stations (250 km with more specific equipment), giving location accuracies of 20 to 40 metres. This precision depends on the number of stations available and their angles respective to the receiver.

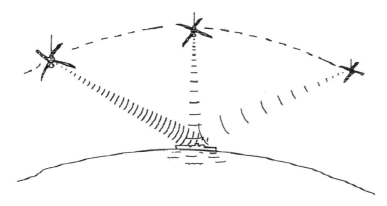

Figure 2.6. Satellite navigation systems (from Telford et al., 1990).

The last few decades have seen the spreading of satellite-based navigation systems. The TRANSIT system, like its Russian counterpart Cicada, is based on a constellation of satellites revolving around the Earth on polar orbits. They transmit continuous waves at 150 and 400 MHz. The measurements of Doppler shifts during the time of observation of each satellite (around 18 minutes) are combined with the speed and heading of the receiving vessel and yield information about the ship's position. Up to 20 observations are possible each day with accuracies of 30 metres. Between these fixes, position is usually computed by dead-reckoning (close estimates of the current positions from the ship's speed and heading), thus degrading the accuracy to a few hundred of metres or more. Other similar satellite-based systems are available (e.g. Doris, Glonass).

These navigation systems all seem to have been superseded by the GPS system commercially released in 1993. The Global Positioning System (GPS) consists of a constellation of 24 satellites in synchronous orbits, associated ground stations and user equipment. Each satellite transmits two codes, the P (Precision) code and the CA (Coarse Acquisition) code. With the CA code, precision is in the order of 100 metres. To gain access to the full precision of the system, a user will need a knowledge of the P code, usually unaccessible to commercial users. This P-code yields positioning accuracies of a few metres, world-wide. At least 3 of the satellites are visible at any moment, which allows near-continuous measurements of 3-D position (latitude, longitude, altitude above the geoid) as well as horizontal and vertical velocities and time (through the satellite's atomic clock). Modern surveys now rely heavily on GPS but still use the other systems as a back-up and in order to cross-verify the location accuracies (see Chapter 3: Sonar Data Processing).

The good resolution of GPS can be enhanced with the Differential GPS (DGPS) scheme. In DGPS, a high-precision GPS reference station receiver is surveyed-in with centimetre accuracy and its position programmed into memory. The reference station now calculates its position using the same satellites as the ship. As it knows its position with a better accuracy, it can identify the satellite signal time travel errors and compute a correction value for each satellite in view. These corrections are regularly updated every second and broadcast to the ship. This technique works because the satellite signal time distortion will be almost the same for any two points on Earth which are less than 500 kilometres apart. The ship now applies these differential corrections and computes its position in real-time with validated 1-m to 5-m accuracy.

Heading and speed are usually measured with 3-D accelerometers, one integration of acceleration giving speed and the next acceleration giving displacement. The speed is also measured with hull-mounted Doppler velocimeters. Accuracies are usually good, although varying from one instrument to the other, and subject to local weather conditions.

2.5.2 Towfish Navigation

In many applications, the sonar package is towed by the ship, sometimes only a few hundred metres above the seafloor (deep-tow). The platform, or towfish, may be positioned accurately by calculating how far back it lies behind and below the ship, or by using acoustic ranging techniques.

The layback of the towfish is computed from the length of cable out and simple trigonometric relations assuming the cable is straight. It is obviously inaccurate for long cable lengths or when the ship is changing speed or heading (and the cable becomes slack). The absolute position of the towfish can be estimated with a precision of a few hundred metres for cable lengths of 4 km and more. This is, however, the easiest and cheapest of the positioning methods. Its inaccuracies can be mitigated by a precise recording of the length of cable out and, when possible, successive runs over the target area or comparison with known reference points (see Chapter 3: Sonar Data Processing). The layback method is most suited to large-scale surveys and along routes.

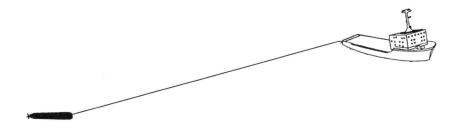

Figure 2.7. Layback of the towfish. The cable is assumed straight.

Acoustic ranging systems are based on two types of configuration: short baseline (SBL) and long baseline (LBL), and variants thereof. The short baseline (SBL) consists in acoustic transducers placed on the hull of the ship as a net, and receiving signals from a transponder on the towfish. The time delays are converted to distances and the relative position of the towfish can be computed with accuracies of 1% for distances out to 2,500 metres.

The long baseline (LBL) consists in acoustic transponders emplaced on the seafloor and accurately positioned. A transponder on the towfish interrogates acoustically the different elements of the net and computes the towfish position by triangulation. The maximum range between transducers in the net is about 3,000 metres.

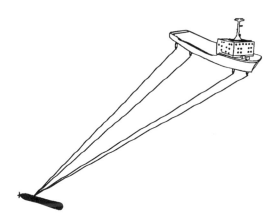

Figure 2.8. Short baseline system (SBL).

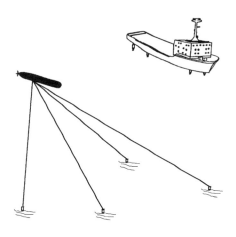

Figure 2.9. Long baseline system (LBL).

Other possible configurations are the ultra-short baseline (USBL), where an array of transducers closely mounted on a pole is attached to the towing vessel to measure the horizontal and vertical angles, the integrated long and short baseline (ILSBL), etc. For very precise positioning, multiple vessels may also be used (this is rare). But the most modern long-baseline nets provide submetric positioning accuracies. Most of the deep-tow platforms have a depth sounder or profiler, which gives information about the height of the vehicle above the seafloor.

In general, precise positioning in depths greater than 1,500 metres is very expensive. It will be required when the desired sonar resolution is high and when the areas to survey are small enough. For larger-scale surveys and surveys along routes, the layback method is the most cost-effective.

2.5.3 Attitude Information

The other important question ("Which area of the seafloor is it ensonifying?") is answered by looking at the attitude of the ship or sonar platform. Four types of movements are recognised: heave, roll, pitch and yaw (Figure 2.10). Heave corresponds to the small-scale variations in altitude of the sonar platform (in the YZ plane). Roll is a lateral movement of the towfish around its longitudinal axis (in the XZ plane). Pitch is a side-to-side movement of the nose and tail of the towfish around its horizontal axis (in the YZ plane). And yaw is a side-to-side movement of the towfish around its vertical axis (in the XY plane). The movements experienced by the ship will not always induce similar movements of the platform (existence of a neutrally buoyant decoupler) and these parameters need to be explicitly recorded for the sonar platform. They are usually measured through gyrocompasses with accuracies of a few degrees to a few tens of degrees. Their influence on sonar images is presented in the next chapters (Chapter 3: Sonar Data Processing; Chapter 9: Image Anomalies and Sonar System Artefacts).

2.6 SUMMARY

This chapter has presented the basic elements of acoustic remote sensing and in particular of sidescan sonar imagery. Although knowledge of them is not mandatory for the interpretation of sonar images, it is most helpful in assessing the validity of the intepretation, avoiding the most common mistakes, and setting confidence levels based on the actual conditions of acquisition. The presentation of the different types of sonar systems available, and of the most modern navigation techniques, should allow the end-user to better understand the requirements for different kinds of surveys. For example, the laying of an off-shore cable or pipeline would require a large swath bathymetric survey to locate the possible route (for example with the Simrad EM-12), and then another survey with a high-resolution sonar (e.g. TOBI) which allows the distinction at a metric scale of possible hazards. GPS navigation and, in the case of towfish layback, repositioning of specific landmarks (see section 3.6.2) with respect to bathymetry should be largely sufficient. Similarly, the search for a shipwreck would require a first survey with a large swath width to locate the approximate wreck site and then a high-resolution survey (e.g. with DSL-120 flown at different heights). The latter would require the precise positioning of the towfish with a short- or long-baseline transponder net. Of course, not all applications require two different surveys. For example, exploration mapping of deep-sea areas such as EEZs is very well attained with one type of sonar (e.g. GLORIA, used in the survey of the US margins; or TOBI, used for mid-ocean ridges). The data produced by these surveys is, however, far from perfectly accurate and trustworthy. Although it may be used as it is in some simple applications, it needs to be radiometrically and geometrically corrected. The next chapter will endeavour to review the different techniques for the processing of raw sidescan sonar data, and producing images that are correct representations of the seafloor and its morphology.

2.7 FURTHER READING

Brekhovskikh, L.M., Yu. P. Lysanov; "Fundamentals of Ocean Acoustics", 2nd edition, Berlin: Springer-Verlag, 270 pp., 1991

Kleinrock, M.C.; "Capabilities of some systems used to survey the deep-sea floor", CRC Handbook of Geophysical Exploration at Sea, Hard Minerals, p. 36-90, R.A. Geyer ed. CRC Press, Boca Raton, 1992

Robertson, K.G.; "Deep Sea Navigation Techniques", Marine Geophysical Researches, vol. 12, p. 3-8, 1990

Telford, W.M., L.P. Geldart, R.E. Sheriff; "Applied Geophysics - Second Edition", 770 pp., Cambridge University Press, UK, 1990

3

Sonar Data Processing

3.1 INTRODUCTION

The exact definitions of the term "sonar processing" vary from one users' community to the other and encompass all possibilities from the display of individual pings' profiles to the production of perfectly accurate maps. To avoid confusion and misinterpretation, we have decided to divide into three distinct stages the transformation of raw sonar data into usable images :

- pre-processing: the preparation of the raw sonar data for processing. This includes but is not limited to the cleaning of the navigation and sensor's attitude files and the conversion between formats.

- processing: processing *per se* is the transformation of the raw swath data to usable images or grids, that will be radiometrically and geometrically correct representations of the seafloor. This includes grid interpolation and mosaicking.

- post-processing: all the operations which are not necessary for a correct interpretation but constitute a definite plus. This includes the computation of statistics, cosmetic operations (e.g. contrast enhancement), the removal of survey-scale noise and multiple reflections. More advanced image analysis techniques are detailed in Chapter 10 (Computer-Assisted Interpretation).

All these stages will be described in their logical order, although not all of them are required during standard operations. At the end of this chapter, two examples of state-of-the-art processing will also be provided, showing what can be routinely performed at sea in near-real time.

3.2 PRE-PROCESSING

3.2.1 Data Formats

Preparation of raw sonar data for processing often includes the conversion between the format delivered by the acquisition system and the format of the processing software. These formats must contain as much as possible of the information needed for processing. During the reassessment of a third-party shipwreck survey, one of the authors recently had to deal with sonar data without any information about the footprint size and frequency, which is altogether annoying. Sonar imagery formats should also be translatable easily from one computer type to the other. Anyone having dealt simultaneously with VAX, IBM and Sun binary data knows the possible nightmares arising from this. In practice, nearly every kind of sonar is recorded into a specific format. Industrial users tend to prefer in-house formats tailored to their particular needs. Academic users are increasingly relying on versatile architecture-independent formats such as HDF (Hierarchical Data Format, a public domain service from the National Center for Supercomputer Applications, USA) and its simplified offspring NetCDF (freely accessible on Internet at *ftp.ncsa.uiuc.edu*). Several international initiatives are currently aiming at a common data format for research imagery (see associated bibliography) and a progressive blending of long-established formats within NetCDF seems the most probable future.

3.2.2 Navigation Data Processing

The previous chapter (Chapter 2: Sonar Data Acquisition) presented the acquisition of navigation and attitude data: 3-D positions, heading, speed, heave, roll, pitch, yaw. Their changes are usually quite smooth, although some spurious values may be introduced during their recording (hardware errors, transmission problems). If undetected or unaccounted for, these variations can lead to important errors during the geometric rectification of the image (see also Chapter 9: Image Anomalies and Sonar System Artefacts). Practical and theoretical studies show that features on the seafloor can be arbitrarily merged, distorted or truncated if navigation and attitude are not correctly processed.

Depending on the navigation system used, ship positions can be inaccurate because of large time intervals between position fixes (for example when no satellite is in view) and when the measurements are excessively noisy (for example because of bad weather conditions, ionospheric perturbations, receiver problems). With the advent of GPS positioning, these inaccuracies are less important than with other systems. They can usually be processed by moving averages which remove the spikes.

Towfish positioning is usually more delicate. When seafloor-emplaced baselines are available, the erroneous values can be related to acquisition hardware problems, fluctuations in the conditions of sound propagation between the transponders, or problems of transmission of this data back to the ship and the recording system. When no baseline is used (layback method), the towfish position is computed from simple

trigonometric relations (Figure 3.1). The horizontal distance is $d = \sqrt{L^2 - (H-h)^2}$, where L is the length of cable out, H is the ship's height above the seafloor and h the towfish's altitude (from its own altimeter). The errors are thus directly related to the errors on the altitudes and the length of cable out. This method is obviously applicable only during the linear portions of the surveys, and not during turns where the across-track displacement becomes important.

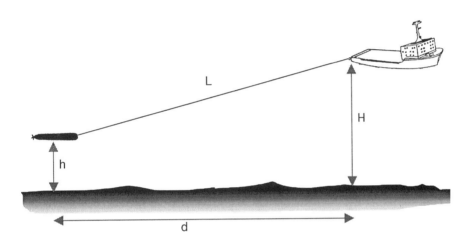

Figure 3.1. Relation between the respective positions of the survey vessel and the towfish.

A new technique in use with TOBI data (LeBas et al., 1994) matches the profiles from the altimeter aboard the sonar platform and the ship's vertical bathymetry measurements. Again, profile-matching only works for linear tracks. During turns, the towfish is deported away and its depth profile needs to be matched with the multibeam bathymetry away from the ship's track. Because all possible combinations must be tested, this proves very computationally intensive and is rarely if ever used.

3.2.3 Attitude Data Processing

For the same reasons (hardware problems), attitude information is also prone to noise and outliers. Along-track profiles typical from a deep-sea survey are shown in Figure 3.2. The towfish's altitude varies regularly, but some spikes of nearly 100 metres are visible here and there. These improbable values are due to the loss of the bottom by the platform's altimeter. The platform's pitch and roll exhibit important variations at the same moments. The roll variations are less important, but have a poor signal-to-noise ratio. If left unfiltered, these variations would create positioning inaccuracies of a few tens of metres on the seafloor (for an initial footprint close to 1 m).

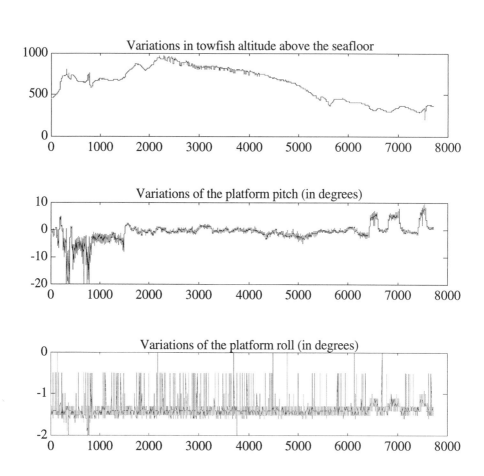

Figure 3.2. Along-track variations of attitude during an actual survey.

Several techniques can be used for the detection and removal of outliers and spikes. Apart from manual selection, they usually rely on smoothing filters (moving average, Kalman filters ...) and interpolation with polynomial or spline functions. Although less regarded, these pre-processing operations are essential to the formation of correct maps of the seafloor. Often, when thorough processing of sonar imagery fails to produce the expected results, another good look at the navigation and attitude helps enhance the images.

3.3 RADIOMETRIC CORRECTIONS

The first stages of processing *per se* are concerned with the recalibration of the individual sonar backscatter measurements. These operations are performed first on the individual measurements (requantisation), second on the successive profiles (across-track), and third on the whole image being formed.

3.3.1 Requantisation

Most sonar processing packages work with measurements rescaled between 0 and 255 (8-bit quantisation). The output from the sonar hardware does not always follow the same quantisation schemes, and it is necessary to resample the data. If not performed correctly, this operation will irretrievably lose the dynamic range of the original backscatter. In extreme cases, the return signals may be saturated or under-saturated for complete regions of the seafloor, thus rendering their analysis impossible. One technique commonly used is the Bell μ-law. Its expression is :

$$\text{new\_value} = C \times Ln. \left(1 + \frac{2^n \times \text{raw\_value}}{2^m} \right)$$

where C is a user-defined constant, m the original number of bits, n the new number of bits (usually n = 8 for a range 0-255). This logarithmic condensing of the data is most useful in low-backscatter areas (abyssal plains, sedimentary fans), because it creates more contrast. In areas already highly contrasted (e.g. mid-oceanic ridges), requantisation through simple division by 2^{m-n} is more adequate. Another advantage of re-quantisation is that it lessens the costs of further computation and storage.

In the past, sidescan sonar imagery was output on electrical line-scan recorders in which high voltages (high signal values) were represented as black. However, the human eye intuitively interprets dark areas as shadows, and bright areas as illuminated. For this reason, sidescan sonar images in this book are represented with highest backscatter values as white and lowest backscatter values as black. This is consistent with the representation scheme used in other domains of remote sensing such as optical and radar imagery.

3.3.2 Across-Track Corrections

The backscattered signal is orders of magnitude lower than the transmitted signal because of beam spreading and attenuation through the water column. For a spherical wave, this attenuation varies as the inverse square of the distance R to the target. It will be different for each returning signal (Figure 3.3). The attenuation of the acoustic beams is multiplied by an additional term in $e^{-\alpha R}$ corresponding to the local conditions of water temperature, salinity and hydrostatic pressure and usually negligible. The local variations make the TVG vary spatially or as a function of time. They are measured by in-situ probes (e.g. XBT) or hull-mounted velocimeters, and these shallow measurements extended down to the seafloor by comparison with databases such as the Levitus database. Sound velocity variations bend the acoustic rays as they go, except for deep-tow sonars for which these variations are negligible. Attenuation is a complex function of the beam width and is determined empirically for each sensor. This is usually performed by surveying a flat featureless portion of the seafloor (such as an abyssal plain). The resulting tables are used to apply a time-varying gain (TVG) to amplify each backscattered return according to its arrival time.

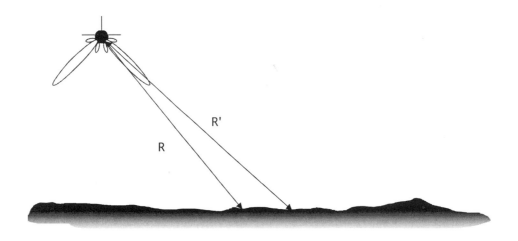

Figure 3.3. Backscatter signals are attenuated differently, depending on the distance they travel.

Many systems have built-in amplifiers that decompose the TVG into several steps. For returns with the same theoretical acoustic properties, the across-track profiles would in fact have discrete steps of amplitudes proportional to the quality of the built-in time-varying gain. A smoother version needs to be recomputed and applied to the relevant points. This also serves for correction when the TVG is not appropriate to the surveyed area. Because of transducer "ageing" or changes in the beam shape, the determination of the angular attenuation coefficients should ideally be performed before each survey. In a different domain, satellites like SPOT routinely calibrate their optical transducers by looking at the Crau plains in the South of France, which are flat, featureless, and with a known reflectivity. Airborne and spaceborne radar missions also make use of specific artificial targets (dihedral for example), whose radar response is known and tabulated.

3.3.3 Along-Track Corrections

Across- track direction

Figure 3.4. Examples of along- and across-track striping.

Systematic radiometric variations are often visible on the sonar images being processed (Figure 3.4). Uniformly black lines (drop-out lines) occur across-track. They are attributable to acquisition problems within the transducer, or loss of data during the transmission between the sonar and the ship. Lines with anomalously low values (across-track striping) may occur due to vehicle motion or various system errors. Some lines may also be shifted across-track, part of the line being black. This last effect is usually attributable to the sonar's altimeter losing the bottom and subsequently timing the backscattered returns inaccurately. Across-track striping is removable through a filtering process which detects low-amplitude lines and replaces the values using the averaged values of adjacent lines. Systematic along-track variations (along-track striping) are due to unaccounted systematic angular variations of the receiving gain. They are corrected by reassessing the time-varing gains or, more easily, by averaging with the along-track profiles computed for each angular bin (angle-varying gain, AVG).

3.4 GEOMETRIC CORRECTIONS

3.4.1 Slant-Range Correction

Raw sidescan sonar imagery presents important across-track geometric distortions, known as slant-range distortions. They occur because sonar systems actually measure the time for a transmitted pulse to travel from the transducer to the target and back to the transducer. Figure 3.5 shows the slant-range distortion: two targets close to the nadir (D_1 and D_2) will be associated with nearly identical slant-ranges R_1 and R_2. Conversely, two targets at far range (D_3 and D_4), at the same distance from one another, will be associated with very different slant-ranges R_3 and R_4 and therefore placed further apart. Without slant-range correction, near-range areas are more compressed than far-range areas. Some specific applications do not use slant-range correction because they need to keep the information just below the sonar's track: fisheries and some target search surveys, but the uncorrected images are not accurate representations of the seafloor.

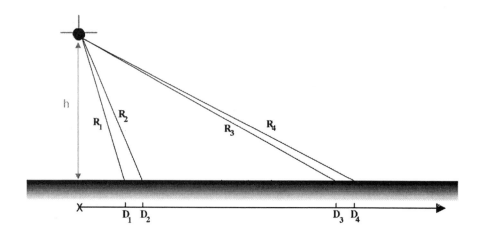

Figure 3.5. Slant-range distortion

Slant-range correction is a remapping of pixels from their apparent position to the true one and is computed from the elapsed time and the sonar platform's height. It should take into account the variation of velocity and of the depth at the point of interest. In practice, neither is done because only the pixels within one or two water depths are remapped by much more than one pixel dimension and in this region the error involved in assuming rectilinear propagation is small. At further ranges the remapping is too small to show the error.

Therefore, assuming a flat seafloor, the correct distance on the ground is:

$$D_i = \sqrt{\left(\frac{c\,T_i}{2}\right)^2 - h^2}$$

where the slant-range distance is $R_i = \frac{c\,T_i}{2}$ and h is the sonar platform's altitude.

Slant-range correction assumes a flat seafloor across-track. This can be a problem in areas of high relief, creating two sorts of artefacts: layover and foreshortening (see Chapter 9: Image Anomalies and Sonar System Artefacts). When bathymetry is available, this should be included in the later stages of processing.

3.4.2 Anamorphosis

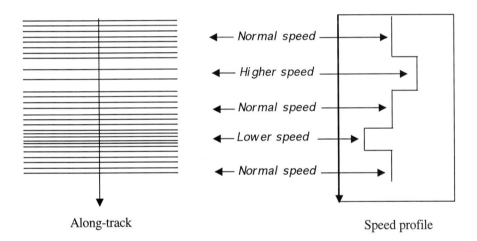

Along-track   Speed profile

Figure 3.6. Variations in the survey speed create distortions in the along-track footprint. These distortions are corrected through anamorphosis.

After slant-range correction, the sonar image is geometrically correct across-track, or as close as it can be. The along-track correction accounts for the variations in platform speed (Figure 3.6). This process, called anamorphosis, produces an image in which the inter-pixel spacing is the same across- and along-track. We saw in Chapter 2 (Sonar Data Acquisition) that the nominal along-track spacing is determined by either the width of the horizontal beam on the ground or the distance travelled by the transducer during the reception interval, whichever is smallest. Successive lines are replicated if the local speed produces a smaller along-track resolution (higher speed), and subsampled if the speed is lower. The anamorphic correction aims at producing an image with a 1:1 aspect ratio, in which geological features are correctly represented.

3.5 BACKSCATTERING MODELS

Sidescan sonar images present as grey levels the local backscattering coefficients, dimensionless numbers also called backscattering cross-sections by analogy with radar remote sensing. Their logarithmic form is often encountered in the literature and is called the backscattering strength. An important number of experiments have been carried and

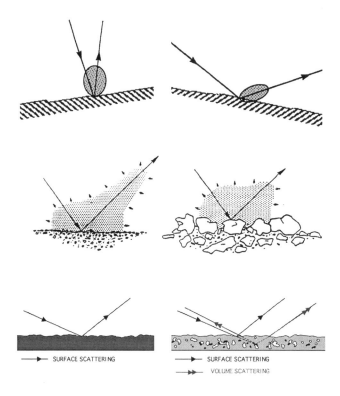

Figure 3.7. Backscattering from the seafloor is influenced by three factors (from top to bottom): local geometry of ensonification, roughness of the seafloor at scales comparable to the sonar's wavelength, intrinsic properties of the seafloor (e.g. rocks vs sediments).

are still carried out to determine the dependence of backscatter on the three factors outlined in Chapter 2 (Sonar Data Acquisition): local geometry of the sensor-target system (angle of incidence of each beam, local slope, etc.), the morphological characteristics of the surface (e.g. micro-scale roughness) and its intrinsic nature (composition, density, relative importance of volume vs. surface diffusion for the selected frequency) (Figure 3.7).

Backscattering models aim at determining which of these three factors governs the seafloor's acoustic behaviour for each couple of {frequency ν; angle of incidence θ}. In radar remote sensing, theoretical models have been fitted to semi-empirical models and compared with radar data acquired with all possible combinations of {ν; θ}. When the Magellan mission started orbiting Venus in 1990, application of these models allowed the geologists to know which effect was predominant in each image (Figure 3.8). For images acquired above latitude 10°N, the angle of incidence is 47° and therefore micro-scale roughness is the predominant effect. Bright pixels in the image correspond to surfaces rougher than the wavelength in use (12.6 cm), dark pixels to smooth surfaces. Conversely, for latitudes below 63°S (i.e. angle of incidence of 19°), local slopes are the predominant factor. Bright pixels correspond to slopes facing the radar, dark pixels to slopes facing away from the radar.

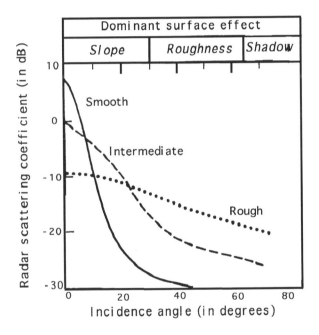

Figure 3.8. Example of a backscattering model. (From Ford et al., 1989)

Unfortunately, no such model exists for acoustic backscatter. The simplest model is the Lambert law, stating that backscatter is a linear function of the square cosine of the incidence angle. Because it is quite simple to understand and does not include heavy mathematics, this model has been quite popular for a time. However, it is far from the complex reality and only works for very simple applications (smooth surfaces and large-scale undulations). Other more complex approaches use the same principle: the sound field scattered by the seafloor consists in elementary waves in mutual phase interference.

Formulations based on the Helmholtz-Kirchhoff theory describe the seafloor by its root-mean-square (rms) roughness in the vertical dimension and a correlation radius in the lateral dimension. Another formulation describes the rough surface as a random distribution of point scatterers independent of each other. Many recent models are using these approaches, refining them semi-empirically or combining them with other techniques (Perturbation Theory, Momentum Transfer Expansion, etc.). A discussion of their relative merits would go too far for the scope of the present Handbook, especially as they are always restricted to certain ranges of frequencies and seafloor types (mainly sedimentary areas). No consensus exists on which model is the optimum one and research in this field is still very active. Regularly, journals such as the *Journal of the Acoustical Society of America*, its Russian counterpart *Acoustical Physics*, or the European *Acoustica* publish new models and results of their applications.

3.6 MAP PRODUCTION

3.6.1 Mosaicking - Stencilling

The processed sidescan sonar imagery is composed of picture elements (pixels) organised into one or several images. These pixels are located by relative coordinates, usually respective to the upper left corner of the image. Georeferencing, also known as geocoding, is the transformation of these relative coordinates into absolute coordinates such as latitude and longitude. It is the first step toward the merging of images and the production of maps. Georeferencing also includes the rectification of the image(s) to a particular map projection (Mercator, UTM, Lambert-Gauss, etc.)

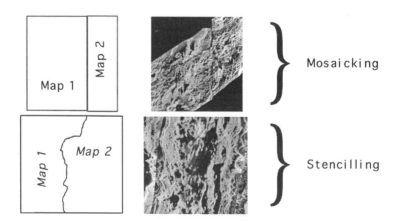

Figure 3.9. Overlapping images can be merged by mosaicking or stencilling.

The merging of different images can be decomposed into two parts: stencilling and mosaicking (Figure 3.9). When the boundary between two overlapping images is linear, or when overlapping pixels values can be averaged, mosaicking merges them into one single dataset. But, when the border between these images is more complex or when averaging of pixel values is to be avoided, the boundary is defined by hand. This is stencilling. Stencilling is particularly useful when two sidescan swaths imaged with opposed directions are overlapping (Figure 3.9).

3.6.2 Interpolation - Rubbersheeting

Several interpolation schemes exist. The simple ones average overlapping pixels with the mean or median values. More complex interpolation methods are available, such as krigeing, polynomial or spline-fitting, or finite element techniques. The accuracy of the interpolation will depend on the original spacing and trends of the points. Discussion of their respective merits are given in reviews such as Lancaster and Salkauskas (1986) or Smith and Wessel (1990). Although they are more needed during bathymetric data processing, interpolation algorithms are very useful when successive swaths are overlapping (Figure 3.10) or when additional geometric corrections are needed.

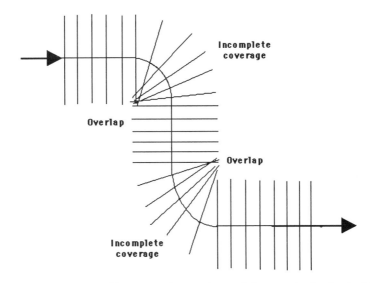

Figure 3.10. When the ship turns, the portions of the swaths in the inside corners are overlapping. The portions of the swaths in the outside corners are more widely spaced and the coverage is incomplete.

The most elaborate interpolation schemes are used to "tie" an image with inaccurate or imprecise positioning to another precisely located image or series of points. This last process, called rubbersheeting, is most often used to co-register a sidescan image to a bathymetric map. For example, TOBI sidescan data is imprecisely located because of the errors on the length of cable out (especially during ship's turns). Recognisable features (e.g. seamounts) are located on TOBI imagery and GPS-positioned bathymetry. They serve as control points to modify the geometry of TOBI imagery so that the control points have the same coordinates. The image on this book's cover were produced this way. Obviously, the number of control points needs to be large enough to provide a resaonable accuracy in the maps thus produced.

3.7 POST-PROCESSING

3.7.1 Image Statistics

The definition and domain of post-processing operations varies quite widely from one author to another. We have therefore chosen to restrict the term "post-processing" to the operations which constitute a definite plus for a correct interpretation, but do not affect significantly the numerical information available in the processed image. Other, more complex, operations that affect this information will be described in Chapter 10 (Computer-Assisted Interpretation).

The numerical information in the image is first described by the first-order statistics, i.e. measures of the relative frequencies of apparition of the different grey levels. The distribution of these frequencies is plotted through the histogram (Figure 3.11). Its shape helps define the data statistics: unimodal (one single peak) or multimodal (several peaks), smooth or irregular, etc. From its limits, the minimum grey level I_{min} and maximum I_{Max} found in the whole image define the dynamic range D :

$$D = (I_{Max} - I_{min})$$

The contrast C is a measurement of the separation of pixel values within the histogram and is defined by :

$$C = \frac{I_{Max} - I_{min}}{I_{Max} + I_{min}}$$

If the grey levels are grouped around a single peak, the image will have a low contrast. Conversely, if the grey levels are spread evenly throughout the histogram, the contrast will be higher. If the image is uniform and only presents one grey level, the contrast is minimal. Associated measures of the grey levels distribution are the mean, the median (most frequently occurring grey level), the standard deviation, and other statistical moments such as skewness or kurtosis.

Their repartition is shown on the TOBI image in Figure 3.11. This image is coded on 8 bits, i.e. with 256 grey levels ranging between 0 and 255. The minimum grey level found in the image is $I_{min} = 54$, the maximum $I_{Max} = 251$. The dynamic range is quite important (D = 197, 77% of the available dynamic range), the contrast as well (C = 0.64). The mean value is 122, the median value 116, and the grey levels are grouped around the mean value with a standard deviation of 51.85. The two principal regions visible in the histogram are the peak of low grey levels corresponding to the shadows (in darker tones in the image), and the smooth central region of freshly erupted basalts.

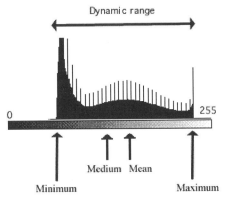

Figure 3.11. The aspect of an image can be described by its statistics.

3.7.2 Histogram Manipulations

Evaluation of an image quality is a very subjective process. Therefore image enhancement is not aiming at increasing the "quality" of the sidescan sonar image, but at increasing the separation between the different regions deemed interesting by the interpreter. Most enhancement operations are mathematical manipulations of the grey levels. Histogram sliding consists in adding (or subtracting) a fixed value to all pixels in the image, hence "sliding" or shifting the histogram from one side of the dynamic range to the other. If the offset is too large, the histogram will be truncated: some grey levels will be forced to the minimum of the range (undersaturation) or to the maximum (saturation).

Similar to histogram sliding, histogram stretching redistributes the pixel values in order to increase the contrast and dynamic range, by multiplication by a constant and rescaling. For an image quantised on 8 bits, each new grey level will be given by :

$$I_{new} = \frac{I_{old} - I_{Max}}{I_{Max} + I_{min}} \times 255$$

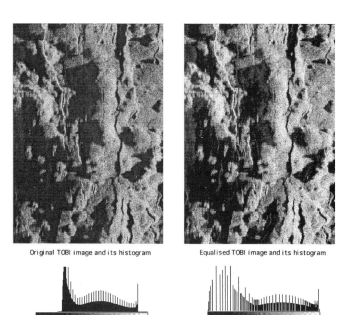

Original TOBI image and its histogram   Equalised TOBI image and its histogram

Figure 3.12. Example of histogram equalisation. The original TOBI image (left) shows a histogram with a high number of dark pixels. The histogram of the equalised image (right) shows increasing contrasts for grey levels which occur frequently, and decreasing contrasts for the other grey levels.

Histogram stretching can also produce grey levels outside of the range available. Its main advantage is to usually improve the appearance of poorly contrasted images with a small dynamic range.

These two operations, sliding and stretching, are usually combined into "histogram equalisation". The idea is to adjust the new pixel assignments to specific grey levels so that there are an equal number of pixels with each new amplitude value. For example, if 64,000 pixels are to be displayed along 64 grey levels, histogram equalisation will partition the range of values in 64 intervals with around 1,000 pixels each. The effect of equalisation is to expand the contrast between grey levels that occur frequently and to decrease the contrast between those pixel amplitudes that occur infrequently, using the limited amount of available image contrast where it is most effective. The mathematical operations necessary to perform image equalisation are described by Gonzalez and Wintz (1977). An example of histogram equalisation is given in Figure 3.12. Histogram equalisation expands the contrast between grey levels that occur frequently and decreases the contrast between grey levels that occur more rarely. This is clearly visible in the two histograms ("before" and "after"). Details are more visible than previously. To further enhance details in specific regions, other histogram manipulations are available: piecewise stretching to stretch independently different portions of the histogram, linear stretching, stretching according to predetermined histogram shapes ...

3.7.3 Speckle Removal

Speckle is a high-frequency noise commonly observed in sidescan sonar imagery (Figure 3.13). It is in fact common to all types of remote sensing using coherent radiations (e.g. radar). After interaction with the seafloor, the acoustic waves are not in phase with each other any more. And positive or destructive interferences may occur, producing anomalously high or low returns. Because speckle is difficult to distinguish from real signals at the limit of resolution of the sonar, it proves hard to remove without affecting significantly the image. Numerous algorithms are available in the literature, from the "quick and dirty" (the mean filter, which averages the speckle into the data but lowers the resolution) to the most elaborate (using simulated annealing algorithms).

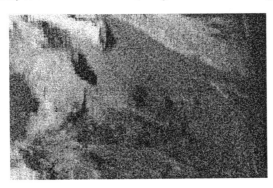

Figure 3.13. Example of speckle (high-frequency noise) in a sonar image.

Common techniques use the local statistics computed on several images, a process known as "multilooking". Local distributions of grey levels are approximated by gaussian or gamma statistics. Mean and variance are computed on small moving windows, and pixels with mean and variance too far from the image mean are considered as speckle and averaged. The results vary from one sensor and one wavelength to another, and no generic despeckling algorithm has been found yet. Moreover, certain applications (e.g. target searches) are looking at the limit of resolution of the sonar, and despeckling can lose valuable data.

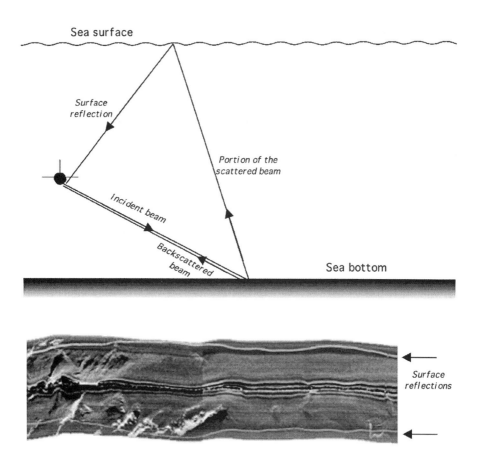

Figure 3.14. Multiple reflections on the sea surface may affect sonar images acquired in shallow waters.

3.7.4 Sea Surface Reflection Removal

Usually in shallow water applications ($\leq$ 1 km depth), sidescan sonar imagery is corrupted by multiple reflections from the sea surface. The first reflection is formed when the sonar beam reflects once from the seafloor and once from the sea surface (Figure 3.14). It manifests itself as bright lines parallel to the sonar track, at a distance from the sonar track roughly equivalent to the water depth.. If the swath is wide enough, subsequent multiples will also be present as equidistant bright lines parallel to the first reflections. They occur principally in sedimentary areas where the seafloor is flat and smooth. Traditional methods attempt to remove the multiple reflections by spatial or frequency filtering. Only one method to our knowledge (LeBas et al., 1994) identifies the multiples explicitly. It is used with TOBI imagery and looks at intervals centred on the across-track distances where the multiple reflections are most likely to occur ((n+1) times the water depth, minus the towfish depth). Searching is conducted within range gates typically 1/10th of the depth wide, centred on this range on each side of the track. Pixel values flagged as surface reflections are replaced by linear interpolation, and random noise added to avoid an overly smooth appearance.

3.8 EXAMPLES OF OPERATIONAL SIDESCAN SONAR PROCESSING

3.8.1 Near Real-Time Processing of TOBI Data

TOBI, a deep-towed sidescan sonar, has been developed by the Southampton Oceanography Centre (ex-Institute of Oceanographic Sciences) in Great Britain, and used in numerous cruises since 1990 (see Chapter 2: Sonar Data Acquisition). TOBI data is processed with homegrown programmes that have been added to the initial WHIPS software provided by Woods Hole. Similar processing is routinely applied to sonar imagery from the shallow-towed GLORIA system. The SOC version of WHIPS is now used for near real-time data processing at sea aboard the research vessel RRS "Charles Darwin". The following description is based on the authors' experience of the system and the article by LeBas et al. (1994).

An estimate of the vehicle position is presently achieved by smoothing the ship's track and using simple trigonometry involving the amount of cable deployed, and extrapolating back behind the ship. Vehicle altitude data is also filtered, using a 280-second median filter. Presently, acoustic imagery is subsampled by a factor of 8, yielding 1,024 points per swath. Occasionally, transmit and receive systems will become asynchronous and miss a whole swath of data, which is unrecorded. Should this occur, a dummy line (duplicated from the previous line) is added to the image to keep the geometry of the image similar.

As the TOBI data is acquired, a time-varying gain is applied within the hardware, to account for attenuation of the signal with increasing range. For a uniformly grey across-track return, this creates small steps in the contouring of recorded imagery. To correct this, a smooth graph of the TVG is created, and compared with the quantised TVG. The difference between the two is applied to the relevant pixels. A geometric correction that needs to be applied to TOBI imagery is the slant-range to ground-range correction. The smoothed altitude of the vehicle is used in the process. The slant-range time is computed for a given horizontal range distance, in order to replicate pixels across the full horizontal range, without gaps forming between pixels. After slant-range correction, each pixel has now a 6-metre across-track resolution. For anamorphosis, variations in ship's speed are accounted and swath lines are replicated or removed according to its value. The final image has a pixel size of 6 m.

Noise present in the data falls into few categories: sharp single pixel peaks, partial swath lines with anomalously low intensity values, dropouts. The former is reduced with a median filter. Dropouts are unfortunately frequently present in TOBI data, and are probably due to the high beam directivity, and maybe loss of signals during transmission through the cable. The average length of a dropout is ~ 1200 m, and only occasionally extends to the full 3000-m swath. It is not generally symmetric about the vehicle's track, and is seen as a single line feature. In some datasets, groups of 2 or more dropouts on successive lines have been experienced. They are replaced by nearest-neighbours interpolation on the lines already checked, in one or two iterations. Deblurring of TOBI imagery is not always performed, as it is quite computationally intensive. Another optional processing is the removal of sea surface reflections occurring in shallow waters.

In the last stage, TOBI imagery is placed on a geographic map, by averaging values corresponding to the projection intervals. Adjacent TOBI swaths are added pass by pass to the final mosaic, trimming the edges of the input imagery should data coincide or overlap in any part of the area. This is done by intervention of a human operator. Checking against the ship-board produced mosaics allows to see the artefacts or the enhancements brought by digital processing. Finally, TOBI data is incorporated into a commercial Geographical Information System package and coregistered with bathymetry.

3.8.2 Processing of High-Resolution DSL-120 Data

Based on the AMS-120 from Klein and Associates, the DSL-120 is a high-resolution sidescan sonar operated by Woods Hole (see Chapter 2: Sonar Data Acquisition). It provides imagery and bathymetry with metric to submetric resolutions. The following description corresponds to the processing as it has been performed at the School of Oceanography, University of Washington (Seattle, USA) and experienced by one of the authors. It uses homegrown software blended with the SONAR package from the Deep-Submergence Laboratory which operates the DSL-120.

Ancillary data include the ship's navigation (GPS positions and dead-reckoning between fixes), the platform's navigation (from a transponder net), and attitude (roll, pitch, yaw, depth, heading). The platform's navigation includes X and Y positions, in metres, relative to the transponder net. It is referenced to geographic coordinates after the processing. Navigation and attitude exhibit some noise and spurious values. Profiles are median filtered. Remaining outliers are removed by the operator and interpolated by cubic splines. Acoustic data (amplitude, phase, time) are recorded by swath lines, along with unprocessed navigation. They are produced at a high rate, producing ~ 1.2 Gb for each pass line, and each swath is approximately 1,024 points wide.

In the course of processing by the SONAR package, amplitudes are corrected for system noise, ambient noise, and cross-talk, by incoherent substraction. Amplitude samples are slant-range corrected, using gridded bathymetry derived from the phase data. It is also possible to use previously processed, and more thoroughly corrected bathymetric grids for slant-range correction. An adaptive angle-varying gain is computed from the average intensity of all samples within discrete angular bins. The angle-varying gain is then inverted and applied to the slant-range corrected amplitude, to compensate for the effects of beam-pattern and average-grazing angle. Individual swaths are processed one after the other, and pixel values in possible overlaps are interpolated. The output pixel size has been selected here as 2 metres. The output files are grids of bathymetry and backscattering amplitude. Larger files are also provided, containing all of the processed data, in the shape of binary triplets (XYZ or XYA), for more detailed processing.

Noise in bathymetry is mainly represented by spurious values and line dropouts. Spurious values are median filtered. Line dropouts, as well as small gaps in the acquisition or between individual swaths, are dealt with during interpolation. Bathymetry is interpolated in 3 or 4 stages, with increasing resolution. At each stage, the original bathymetry is appended to the interpolated one, in order to ensure convergence. In amplitude images, noise is related to speckle and line-dropouts. It may be interesting to note that, because of the tow depth, no sea-surface returns are to be observed. Line-dropouts do not occur systematically in the image, maybe because the attitude parameters (roll, pitch, yaw) are recorded at a high frequency. At the current stage of processing, no need has appeared to correct the line dropouts, but no doubt it will be in the future. Because of the high resolution, speckle is not always distinguishable from significant textural patterns, and is not filtered.

Relocation of DSL-120 data into a geographical frame is made by simple computation of the metric distance from the transponders' net reference, and translation into geographical coordinates. In the case of overlapping passes, bathymetry acquired during each pass is interpolated. The problem is more complicated in the case of amplitude, as drastic trimming would miss the difference in illumination directions, and complicate the possible interpretations. Fortunately, it has not happened so far, because the different passes were adequately spaced. Throughout the processing and post-processing processes, images and grids are checked with ship-board produced mosaics, and additional information such as that obtained from submersible dives.

3.9 SUMMARY

In the course of this chapter, we have endeavoured to present the three different stages of a comprehensive processing of sidescan sonar imagery. This processing has been decomposed in its logical order (Figure 3.15) : pre-processing of navigation and attitude data, radiometric corrections (requantisation, across-track corrections [TVG], along-track corrections [striping, line dropouts, AVG]), and geometric corrections (across-track [slant-range] and along-track [anamorphosis]). A rapid review of backscattering models has been included there to show the conditions influencing the local backscatter. This serves as a guide to the next section about map production (mosaicking, interpolation, rubbersheeting). Finally, the post-processing stages are detailed that enhance the data presentation without affecting it significantly. To illustrate them, practical examples of processing are provided for sidescan sonar imagery with different degrees of resolution (GLORIA: 60 m; TOBI: 6 m; DSL-120: ≤ 1 m). As far as possible, heavy mathematics have been avoided, but all concepts necessary for an accurate interpretation of the sonar images presented in the next chapter have been introduced.

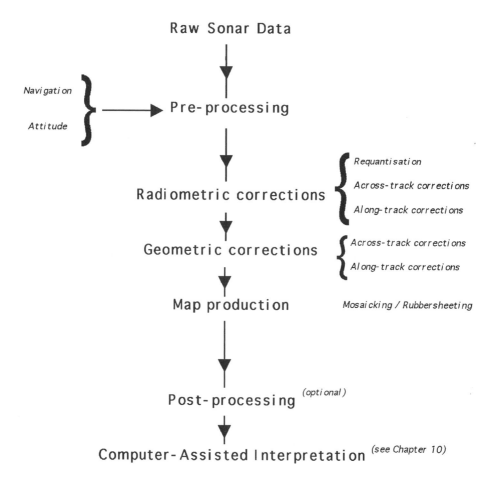

Figure 3.15. Data processing chain.

3.10 FURTHER READING

• **About sonar data formats**

Stewart, W.K.S.; "Subsea data processing standards", RIDGE Workshop Report, 1990

Blondel, Ph., L.M. Parson; "Sonar processing in the UK", BRIDGE Workshop Report no. 5, 14 pp., Jan. 1995

Blondel, Ph., L.M. Parson; "Sonar processing in the U.K.: A short review of existing potential and new developments for the BRIDGE community", BRIDGE Publication no. 1, 27 pp., May 1994

• **About sonar processing**

Fleming B.W.; "Sidescan sonar: a practical guide", Int. Hyd. Rev., vol. LIII, no. 1, p. 65-92, January 1976

Kleinrock, M.C.; "Capabilities of some systems used to survey the deep-sea floor", CRC Handbook of Geophysical Exploration at Sea, Hard Minerals, p. 36-90, R.A. Geyer (ed.), CRC Press, Boca Raton, 1992

LeBas, T.P., D.C. Mason, N.W. Millard; "TOBI image processing: The state of the art", IEEE J. Oceanic Eng., vol. 20, no. 1, p. 85-93, 1995

Mason, D.C., T.P. LeBas, I. Sewell, C. Angelikaki; "Deblurring of GLORIA sidescan sonar images", Marine Geophys. Res., vol. 14, no. 2, p. 125-136, 1992

de Moustier, C.; "State of the art in swath bathymetry survey systems". Int. Hydr. Rev., vol. 65, no. 2, p. 25-54, 1988

Reed,T.B., D. Hussong; "Digital image processing techniques for enhancement and classification of SeaMARC II side scan sonar imagery", J. Geophys. Res., vol. 94, no. B6, p. 7469-7490, 1989

Searle, R.C., T.P. LeBas, N.C. Mitchell, M.L. Somers, L.M. Parson, Ph. Patriat, "GLORIA image processing: the state of the art", Marine Geophys. Res., vol. 12, p. 21-39, 1990

• **About backscattering models**

Boyle, F.A., N.P. Chatiros; "A model for acoustic backscatter from muddy sediments", Journal of the Acoustical Society of America, vol. 98, no. 1, p. 525-530, July 1995

Caruthers, J.W., J.C. Novarini; "Estimating geomorphology and setting the scale partition with a composite-roughness scattering model", IEEE Oceans '93 Proc., vol. III, p. 220-228, 1993

Essen, H.H.; "Perturbation theory applied to sound scattering from a rough sea-floor", SACLANTCEN Report SR-194, 21 pp., 1992

Jackson, D.R., D.P. Winebrenner, A. Ishimaru; "Application of the composite-roughness model to high-frequency bottom backscattering", J. Acoust. Soc. Am., vol. 79, no. 5, p. 1410-1422, 1986

Jackson, D.R., and K. B. Briggs; "High-frequency bottom backscattering: roughness versus sediment volume scattering", J. Acoust. Soc. Am., vol. 92, n. 2, p. 962-977, 1992

Somers, M.L.; "Sonar imaging of the seabed", in Acoustic Signal Processing for Ocean Exploration, Moura and Lourtie eds., p. 355-369, 1993

• About radar remote sensing

Ford, J.P., R.G. Blom, J.A. Crisp, C. Elachi, T.G. Farr, R.S. Saunders, E.E. Theilig, S.D. Wall, S.B. Yewell; "Spaceborne radar observations: a guide for Magellan radar-image analysis", California Institute of Technology, Jet Propulsion Laboratory, Publication 89-41, 132 pp. (NASA-CR-184998), 1989

• About interpolation techniques

Lancaster, P., K. Salkaukas; "Curve and surface fitting: an introduction", Academic Press, London, 280 pp., 1986

Smith, W.H.F., P. Wessel; "Gridding with continuous curvature spline in tension", Geophysics, vol. 55, p. 293-305, 1990

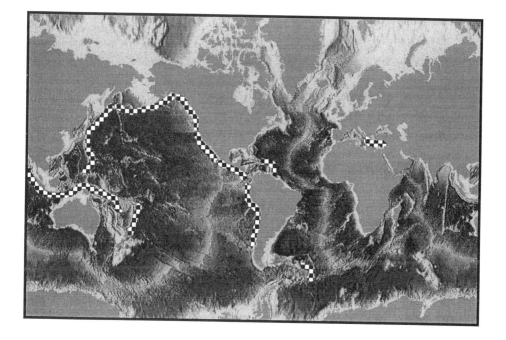

Figure 4.1. Location of the world's deep-ocean trenches and collision margins.

4

Deep-Ocean Trenches and Collision Margins

This chapter is the first actually dealing with examples of sidescan sonar imagery from the different environments defined in Chapter 1. To understand the occurrence, distribution and partitioning of these environments, it is necessary to consider the basic geological processes that shape the Earth. In essence, the planet Earth is a dynamic system. Convection, flow and turbulence are familiar concepts for the atmosphere and hydrosphere. These processes are less obvious, yet just as fundamental, for the interior of the Earth. The rigid plates covering the planet are driven into motion by movements deep into the planet's mantle (Figure 4.2). It is this motion which causes the phenomenon of continental drift and global plate tectonics. In turn, global plate tectonics is the primary process that determines the seafloor environments described in this book. A basic understanding of plate tectonics is therefore necessary to put these environments into context and interpret their constitutive features.

4.1 PLATE TECTONICS AND THE SEAFLOOR ENVIRONMENTS

It is now widely accepted that plate tectonics is the product of the giant heat engine formed by the core and mantle of the Earth's interior (Figure 4.2). The constant convection and overturning of half-solid, half-molten rock, forces the planet's giant crustal plates to creep across the surface. Sometimes, this motion splits and separates the plates. When this happens, the underlying molten rock wells up to fill the gaps, forming the long chains of volcanoes of the mid-ocean ridges (see Chapter 5: Mid-Ocean Ridge Environments). In other places, the plates are colliding with oneanother with such power that one plate slides beneath the other (this is the subduction process), forming deep troughs, the ocean trenches (described in this chapter). In some cases, hot blobs of mantle come up from deep within the Earth's interior and burn through the oceanic crustal plates to erupt as seamount volcanoes (like Hawaii).

Between the ocean trenches and the mid-ocean ridges, the deep ocean floor is, geologically speaking, virtually inactive. Yet there the constant rain of pelagic sediment forms an ever-thickening blanket of mud. Occasionally, these abyssal plains are swept by benthic storms that can last from a week to a month. Episodically, large volumes of sediments are deposited from the neighbouring continental margin in the form of turbidite flows (see Chapter 6: Abyssal Plains and Basins). At the edges of the oceans, where the deep seafloor meets the continents, pelagic rain is also frequently interrupted by giant avalanches of mud and sand. These are the continental rises where river sediments build up into unstable mounds or fans, that collapse when disturbed by earthquakes or even minor sea level changes. These catastrophic episodes erode deep canyons on the sides of the continental rises, and transport debris far out into the abyssal regions where they are deposited as blankets of sand and mud. At the top of the continental rise, where the sea shallows toward the land, lies the continental shelf, with gentle slopes and average depths of 120 m, and down to 250 m. Its width depends on the original geometry of the plate break-up and present-day plate tectonics, and ranges from a few kilometres (adjacent to transform and actively subducting margins) to hundreds of kilometres (at passive orthogonal margins). Bathymetric relief on the continental shelf is mainly caused by reefs, sand and gravel structures, and erosional channels. It is the most thoroughly studied area of the ocean floor (see Chapter 7:

Continental Margins), largely because of its great economic importance in terms of hydrocarbons and fish reserves as well as ecological issues.

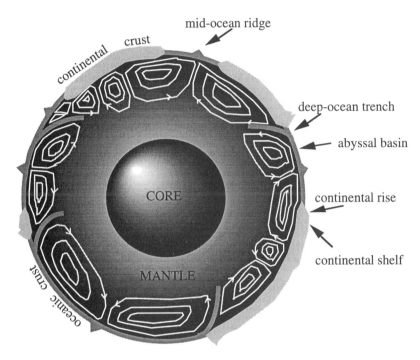

Figure 4.2. Schematic diagram of a section through the Earth (not to scale) showing the relationship between the crustal plates, their margins and the underlying mantle flow that constitutes the process of plate tectonics.

4.2 GEOLOGICAL BACKGROUND

Deep-ocean trenches are the places where the Earth's oceanic crust is destroyed and are the deepest marine environments. Characterised by extreme topographic relief, the deep-ocean trenches are dominated by tectonic and sedimentary processes. As the oceanic crust creeps away from the mid-ocean ridges, it ages, cools and becomes covered in an ever-thickening blanket of sediment. For the oldest ocean crust, around the northern and western rim of the Pacific Ocean, it is so old, cold and dense that it sinks under its own weight back into the Earth's interior. Around the eastern Pacific Ocean's rim, the convective processes from within the Earth's interior force the oceanic crust to dive beneath the American continent. The result of this convergence of the plates is the formation of deep trenches running along the length of the plate margins. The Challenger Deep is some 11,000 metres deep and lies at the bottom of the Marianas Trench. It is the deepest place on the planet's surface. The only vehicle to have penetrated the depths of the Challenger Deep was Picard's *Nautilus* in the 1950's.

These trenches, known as subduction zones, are the focus of the planet's strongest earthquakes. The shocks can be traced for over 700 km into the Earth's interior, mapping the descending ocean plate to its fate. Such earthquakes are often responsible for tsunamis, "tidal" waves that spread across the oceans at the speed of sound, causing havoc and destruction to coastal areas up to thousands of miles away.

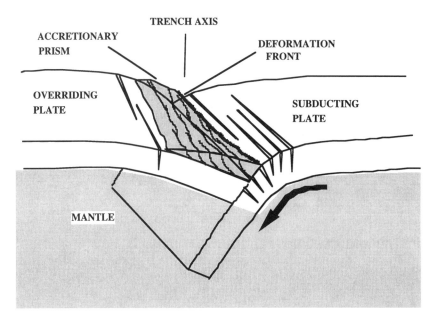

Figure 4.3. Schematic diagram showing a perspective view of an oceanic subduction zone. The ocean crust on the right moves to the left and is subducted beneath the overriding plate on the left.

On the sides of the trenches, the ocean floor is broken-up by faults (Figure 4.3) that often result in landslides. In places, extinct volcanoes and ridges on the subducting crust plough into a trench causing first uplift and then slumping and sliding of enormous piles of sediments on the inner trench wall. This "bulldozer" effect piles up great wedges of wet mud and sand on the overriding side of the trench. Reaching several miles thick, these piles of wet sediment are compressed by their own weight, forcing pressurised water and mud to the surface as mud volcanoes (Dickinson and Seely,

A major effect of the subduction zones is the formation of volcanic chains, called island arcs, in the overriding plate. At 110 kilometres beneath the seafloor, the subducting plate begins to melt. This in turn melts the interior of the Earth and together the molten materials rise to the seafloor and form volcanoes. Such volcanoes are amongst the most destructive on the planet and include examples such as Krakatoa, Santorini, and Mt. St. Helens. But the islands are just the tips of those volcanoes that jut above sea level. Beneath the sea surface there are thousands of these volcanoes, many of which have hydrothermal systems which, unlike their counterparts on the mid-ocean ridges, are often rich in precious and non-ferrous metals. Another form of volcano, formed in the trench, is unique to the subduction zones. These are "cold" volcanoes, erupting a mineral mud called serpentine. Formed when sea water penetrates through the cracks in the walls of the trenches and reacts with the mantle rock deep within the Earth's interior, the serpentine expands and forces its way to the surface (Fryer and Fryer, 1977). It pours onto the seafloor like lava, building-up the flanks of serpentinite seamounts. With the serpentine come cold, super-alkalic fluids deposit minerals on the seafloor as chimneys; a cold analogy of the black smokers.

Investigation of these deep-ocean trenches poses serious technological problems. Mainly, their great depth, their extreme bathymetric variation and, to a lesser extent, their narrowness make investigation by sonar difficult. Surface-towed sonar devices require the power to receive echoes with acceptable signal-to-noise ratios from a vertical two-way travel path that can be as much as 20 km. For sidescan sonars, the great vertical depth results in a wide nadir zone from which little usable data are returned. This is complicated by the narrowness of the deep-ocean trenches. Another complication arises from the long, steep slopes associated with the trench walls that results in poor estimates for slant-range correction. The great depth of the trenches, typically more than 6,000 m, precludes virtually all deep-towed sonar vehicles from these extreme environments.

4.3 THE REASONING BEHIND THE INTERPRETATION

Because this is the first chapter dealing with actual examples of sidescan sonar imagery from a specific geological environment, it is a prudent point at which to introduce a set of criteria behind which lies the methodology of sidescan sonar image interpretation. Before attempting any interpretation of sidescan sonar imagery, it is important to know a number of details about the acquisition system, the survey method and the way the data were processed. When trying to identify specific features within sonar images, it is important to know a number of key parameters: resolution of system, range of targets, depth beneath the sonar, geological context (i.e. deep-ocean, continental shelf, etc.).

For many who are experienced in the field of sidescan sonar interpretation, these parameters are automatically taken into consideration. However, to aid the processes of interpretation, or at least an understanding of how to make an interpretation, we have compiled a check-list of parameters that need to be considered. By repeatedly referring to this check-list when confronted with new sidescan sonar images, the methodology behind the interpretations should become clear, and users will gain confidence in making their own interpretations.

+---+
| **INTERPRETATION CHECKLIST** |
| |
| <u>**For the whole image:**</u> |
| |
| • How was it acquired ? |
| |
| • Type of sonar |
| • Frequency |
| • Configuration (hull-mounted, deep-towed, altitude above the seafloor ...) |
| • Beam width, swath width, resolution |
| • Vehicle information: heading, speed, attitude (roll, pitch, yaw) |
| |
| • How was it processed ? |
| |
| • Analogue or digital processing ? |
| • Time-varying gain ? Angle-varying gain ? |
| • Slant-range correction ? |
| • Anamorphosis ? |
| • Additional processing ? |
| |
| <u>**For each identifiable region:**</u> |
| |
| • What was the illumination geometry for this particular region ? |
| |
| • Range (far, close, intermediate ?) |
| • Angle (sub-perpendicular, grazing, varying ... ?) |
| • Is the region viewed obliquely, i.e. at several angles and ranges ? |
| |
| • What is the backscatter ? |
| • Relative values (high or low) |
| • Variations inside the region (homogeneous or not, contrasted or not ...) |
| |
| • What are the characteristic textures ? |
| • Appearance (smooth, grainy, mixed ...) |
| • Organisation (random, regular patterns, directionality ...) |
| |
| • What are the dimensions of the region ? |
| |
| • What is the relation of the region to its surroundings ? |
| |
| • What can it be ? (extraction of the 3-D object from the 2-D sonar image) |
+---+

Table 4.1. Sidescan sonar interpretation checklist.

HOW? How was the image acquired? Most important is the type of sonar, and the frequency it transmits and receives. As emphasised in previous chapters (Chapter 2: Sonar Data Acquisition; Chapter 3: Sonar Data Processing), the response of the seafloor will differ with the frequency. A surface smooth at low frequency (i.e. long wavelength) may appear rougher at a higher frequency (i.e. shorter wavelength).

For many who are experienced in the field of sidescan sonar interpretation, these parameters are automatically taken into consideration. However, to aid the processes of interpretation, or at least an understanding of how to make an interpretation, we have compiled a check-list of parameters that need to be considered. By repeatedly referring to this check-list when confronted with new sidescan sonar images, the methodology behind the interpretations should become clear, and users will gain confidence in making their own interpretations.

HOW? How was the image acquired? Most important is the type of sonar, and the frequency it transmits and receives. As emphasised in previous chapters (Chapter 2: Sonar Data Acquisition; Chapter 3: Sonar Data Processing), the response of the seafloor will differ with the frequency. A surface smooth at low frequency (i.e. long wavelength) may appear rougher at a higher frequency (i.e. shorter wavelength). Equally important is the knowledge of the configuration in which the sonar was used. Was it hull-mounted or deep-towed? How high was it above the seafloor? Information about the beam width(s) and swath width will prove most useful when looking at the ground resolution. Whatever the sonar platform is, one must also know precisely its heading and speed variations (were there any sudden accelerations? was the platform zigzagging or going straight?), and the variations in attitude (roll, pitch, and yaw). As mentioned earlier (Chapter 3: Sonar Data Processing), unprocessed variations in attitude have been shown to drastically change the shape of objects on the seafloor, or even separate them into what appears to be distinct pieces.

HOW? How was the image processed? Was the image we are looking at processed in an analogue fashion, or digitally? The potential application of time-varying gains (TVG) yields to varying backscatter levels for the same types of objects or processes along the survey line. Conversely, angle-varying gains (AVG) would create across-track variations of backscatter. Another crucial point to consider is the presence or absence of slant-range correction and anamorphosis, because they influence the shapes of each identifiable object and their closeness to the ground reality.

WHAT? What are the characteristics of each of the regions of interest inside the image? And first, what was the imaging geometry? Each morphological region or object will appear differently whether it was imaged at near range or far range, with a sub-perpendicular sonar beam or at low grazing angle, or if both ranges and angles are varying across the region.

WHAT? What does the acoustic backscatter look like for the region of interest? Are the values relatively low or high, homogeneous or heterogeneous, contrasted or not? What is the texture of the target (smooth or grainy or mixed)? Are there any organised patterns inside the region, or is the texture random?

WHAT? What is the relation of the region/object with its surroundings? This does not only concern the dimensions of the region, but also includes the connection with the regions/objects around it, and the extraction of its 3-D aspect out of the 2-D sonar image.

By answering these questions one by one, the interpreter should be able to restrict the hypotheses to only a few, and correctly identify the different features in the sonar image. A reasonable knowledge of the region surveyed, its context, and of what to expect, is of course helpful to the final interpretation. The next chapters will show the application of these principles to sonar imagery taken in all marine environments with as wide a range of instruments as possible.

4.4 DEFORMATION FRONTS AND ACCRETIONARY PRISMS

There are essentially two types of deep-ocean trench, those that are filled with sediment and those that are not. Typically the trenches that form adjacent to continental slopes have a ready supply of sediment that accumulates on the seafloor. The leading edge of the overriding plate of sediment filled trenches is often characterised by a deep pile of material that is known as an accretionary prism (Leggett, 1982). The junction between the two plates is defined by a zone of compressional tectonics, the deformation front, that trends approximately parallel to the axis of the trench.

Figure 4.4 shows a GLORIA image mosaic of the Columbian Trench. GLORIA is a 6.5-kHz shallow-towed long-range sidescan sonar with a working resolution of about 50 x 50 m in the mid-field (see Chapter 2). The most striking feature of the image is the large area of smooth low-backscatter. This represents an area of thick, soft sedimentary fill. Brighter areas are either faults, folds or regions of coarser and rougher sedimentary areas. The bright curvilinear reflector that winds NW-SE across the centre of the scene is the deformation front. This structure is hosted within the trench fill-sediments and delineates the easternmost, or seaward, limit of deformation (Westbrook et al., 1995). The sediments are folded and broken into a series of approximately trench-parallel linear ridges and fault scarps. The deformation front comprises a combination of fold ridges and faults. Landward of the deformation front, the linear features are more folded ridges and fault scarps that both face towards and away from the sidescan sonar. In the top right-hand corner, a number of thin bright linear reflectors trending N-S are downslope channels feeding the trench with sediment. Towards the upper-right hand side of the image, a series of thin curvilinear reflectors and shadows comprise a set of faults and ridges that mark the position of a major strike-slip fault system (Figure 4.6).

An interpretation of the GLORIA mosaic is shown in Figure 4.5. The curvilinear line with barbs pointing towards the east denotes the position of the deformation front. The barbs show the direction of underthrusting, i.e. the direction taken by the down-going plate relative to the upper plate. The hatched area behind the barbed line is the area of the accretionary complex affected by faulting and folding. The curvilinear lines and large arrows show the strike-slip fault zone. In the top right corner, the arrows reveal the direction of downslope sediment transport through the channels into the trench. A few seamounts can be identified on the down-going plate, despite it being covered by a relatively thick sedimentary blanket. This indicates that the volcanoes must be quite prominent or are the result of more recent volcanic activity, i.e. they are younger than the crust upon which they are built.

Deformation of the sedimentary pile leads to fold and slump structures, as well as fault scarps. These are typically aligned parallel to the axis of the trench. In the overriding plate these are often of a compressional nature. On the down-going plate the faults are usually extensional and form scarps that face into the trench. At some subduction systems the plates converge obliquely. In these cases the fault structures are oblique to the axis of the trench. Where the obliquity of convergence is extreme, strike-slip faults develop which can translate fragments of crust laterally.

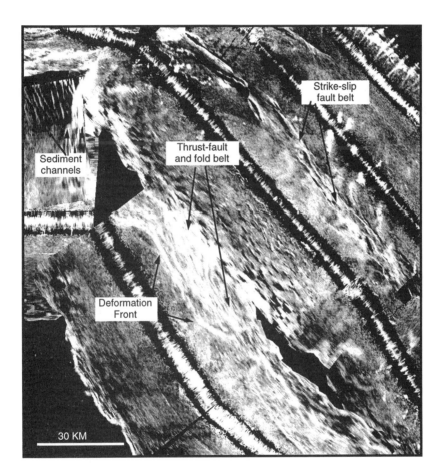

Figure 4.4. GLORIA image mosaic of the Columbian Trench (courtesy of Prof. G. Westbrook, University of Birmingham, UK). The bright curvilinear reflector that winds NW-SE across the centre of the scene is the deformation front and delineates the easternmost, or seaward, limit of deformation.

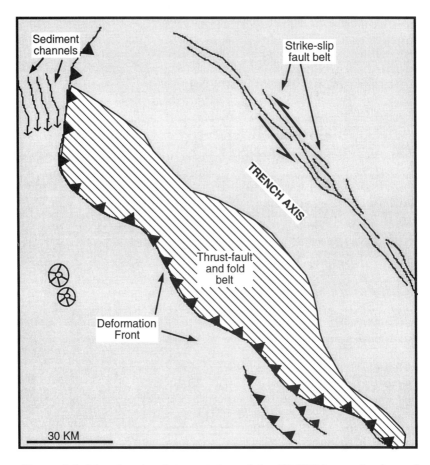

Figure 4.5. Line-drawing interpretation of the GLORIA mosaic shown in Figure 4.4. The barbed curvilinear line denotes the position of the deformation front with the barbs pointing in the direction of underthrusting.

Figure 4.6 shows a detail of the GLORIA sidescan sonar image of the floor of the Columbian Trench, off the north-west coast of South America. A strike-slip fault zone is shown, with the features of interest being the curvilinear fabric that trends left to right across the centre of the image. Linear features that are graded bright to dark in backscatter strength are longitudinal ridges caused by folding of the sediments by compressional forces acting obliquely across the strike-slip zone. These features are characteristic of the formation of "flower structures" that are typically formed during strike-slip displacement, especially within sedimentary sequences (Woodcock and Fischer, 1986; An and Sammis, 1996). The compressional forces are the result of bends in the trace of the fault strands. The sharply defined shadows and thin bright linear features are fault scarps both facing towards and away from the sonar. Also present are small basins, called tear-aparts or relay zones.

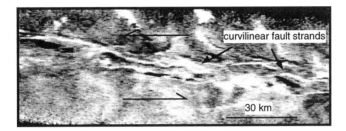

Figure 4.6. GLORIA sidescan sonar image of the floor of the Columbian Trench, off the north-west coast of South America (detail from Figure 4.4) showing the curvilinear fabric of a major strike-slip zone, courtesy of Prof. G. Westbrook, University of Birmingham, UK.

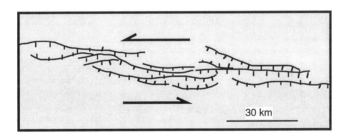

Figure 4.7. Interpretation of the strike-slip fault structures that form curvilinear ridges, scarps and hollows on the floor of the Columbian Trench. The ticks on the longer lines in the interpretation depict facing directions of slopes on the sides of pressure ridges, or fault scarps, caused by compression or extension across the strike-slip fault zone.

Where there is a strong supply of sediment into a trench, or where the colliding plates have thick sedimentary covers, the subduction system piles up these sediment to form an accretionary prism. Often the compressional forces within these prisms of sediment result in fold structures that are often related to the compressional fault structures described above.

High-resolution sidescan sonar imagery (Figure 4.8), made with the IZANAGI shallow-towed system, shows a part of the Nankai Trough located in the northeast of the Philippine Sea (Pickering et al., 1995). The image shows the deformation front to the seaward side of the accretionary prism. Behind this front, the sediment pile is cut by thrusts and folded into a chaotic pattern of ridges and troughs. Bright linear and curvilinear reflectors trending obliquely across the image are folds and faults facing the sonar. Although the thrusts and fold axes are approximately parallel to the axis of the Nankai Trough, the south-eastward convex nature of these features is evidence of overthrusting towards the southeast (Moore et al., 1990). This is consistent with the polarity and direction of plate motion at this subduction system. Smooth low reflectivity material within the axis of the Nankai Trough is derived from soft open-ocean, pelagic, sediments that blanket the surface of the down-going oceanic plate.

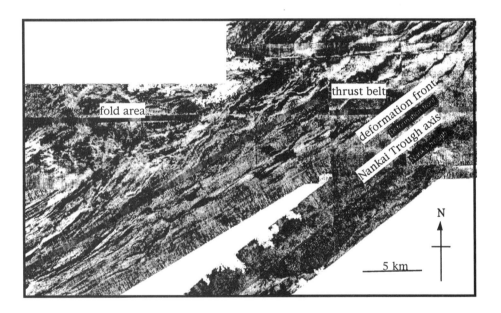

Figure 4.8. IZANAGI high-resolution sidescan sonar image of part of the Nankai Trough (Pickering et al., 1995) showing the deformation front to the seaward side of the accretionary prism. Copyright Chapman & Hall, London, 1995.

In the line-drawing interpretation of the IZANAGI high-resolution sidescan sonar image of the Nankai Trough (Figure 4.9). The trough is filled with sediments and the curvilinear reflector that indicates the position of the deformation front is shown by the barbed line. Behind this deformation front, the sediment pile is cut by thrusts, also shown by barbed lines. For all thrusts and the deformation front, the barbs mark the upper, over-thrust, or hanging-wall portion of the fault. Interestingly, the area of fine striations in the middle of the left-hand side of the scene is evidence of minor fault structures. These also trend parallel to the axis of the Nankai Trough. They are represented by the area of cross-hatching. Behind the thrust and fault belt the more chaotic area of folding is represented by the chevron-hatched area. The origin of these folds may be compression, or slumping, of the accretionary prism.

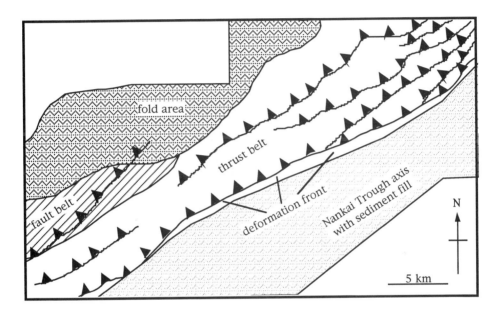

Figure 4.9. Line-drawing interpretation of the IZANAGI high-resolution sidescan sonar image of the sediment-filled Nankai Trough: the deformation front and other thrust faults are shown by barbed lines with the barbs marking the over-thrust portion of the fault.

4.5 TRENCH-FILL STRUCTURES AND PROCESSES

The supply of sediments into the trench is often a characteristic of those trenches formed adjacent to a continental margin. For example, the western margin of South America, especially adjacent to the coast of Chile, supplies material into the Peru-Chile trench. As with other continental margin sedimentary processes, these trench supply systems are characterised by distribution fans, slumps and occasionally contourite deposits (Schweller and Klum, 1978). A major difference between these processes and those on passive continental margins (described in Chapter 7) is the effect of the steeper slopes and frequent earthquakes at destructive plate margins.

Figure 4.10 shows a GLORIA mosaic of the Columbian Trench. The inner trench wall forms the margin of the coast of Columbia and slopes oceanward to a depth of over 6000 m. Sediment supply from the coast forms branching channels that feed into the axis of the trench. The bright tree-like structures on the sidescan sonar image are the channels with the branches near the coast as the minor tributaries and the trunks as the major channels. In the axis of the trench, the bright near-circular areas of reflectance are fans at the ends of the sediment feeders. The cause of the bright reflectivity of these features probably reflects a coarser sediment supply. Lower reflective, smooth areas are finer-grained pelagic sediments forming the accretionary complex. The chaotic reflectors closest to the coast are reflections from the coastal shelf. Refraction of the far-range sonar signal at the base of the thermocline, in the uppermost water column, allows propagation of the sonar wave to these shallow regions.

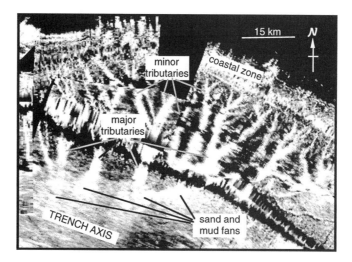

Figure 4.10. GLORIA mosaic of the northern end of the Columbian Trench (courtesy of Prof. G. Westbrook, University of Birmingham, UK).

In the interpretation of the GLORIA image (Figure 4.11), the mid-grey tones show the position of the pelagic sediment pile that forms the accretionary prism. The black-filled areas are the parts of the sediment channels that have been imaged, while the dark stippled areas are those parts of the channels that are inferred to cross the nadir region of the sonar image where there are no data. The mid-grey stipple denotes the sediment fans at the base of the slope. Coastal reflectors are indicated by the coarse black dots on the white background.

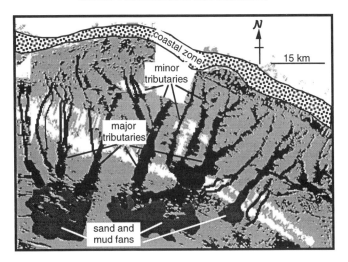

Figure 4.11. An interpretation of the GLORIA image above, showing the major structures of this part of the Columbian Trench.

Features unique to collision margins and deep-ocean trenches are subducting seamounts and other bathymetric features that form part of the down-going plate, e.g. aseismic or active ridges (Fitch, 1972). The effect of such features can be an increase in faulting and a narrowing of the width of the trench system by a process called forearc erosion, e.g. for the Tonga Trench in the West Pacific. The effect of subducting prominent features increases with increasing sediment thicknesses within trenches.

One of the best-recorded examples of this process comes from a GLORIA study of the Chile Trench, where the actively spreading Chile Rise, an oceanic spreading ridge, is subducted beneath the South American Plate (Cande et al., 1987). Figure 4.12 shows a GLORIA mosaic of the collision point between the Chile Rise and the Chile Trench.

The broad swath of bright reflective seafloor, trending northeast-southwest in the upper half of the image is the axis of the actively spreading Chile Rise. Because it comprises fresh lavas, the Chile Rise is relatively rough compared with the surrounding pelagic sediment covered seafloor. This roughness results in stronger backscatter of the sonar signals. In the centre of the upper portion of the image there is a triangular region of mottled chaotic bright and dark patchy seafloor. This indicates an area of sediment folding and slumping, and is the accretionary prism. To the south of the image, this accretionary prism is much reduced in width. The curvilinear bright reflector that winds from the centre of the bottom of the image to the centre-left of the top of the image is the west-facing slope of the deformation front. Bright and dark linear features around the southern end of the ridge axis are reactivated spreading faults. By comparison with the ensonification direction, it can be seen that these faults face west.

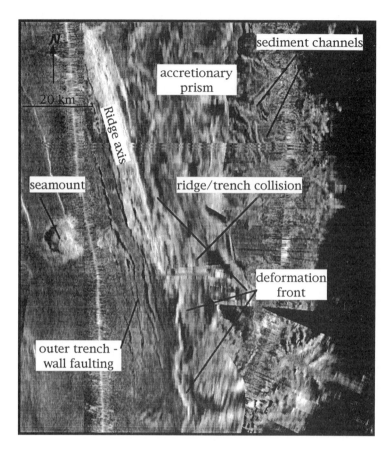

Figure 4.12. GLORIA mosaic of the collision point between the Chile Rise and the Chile Trench, courtesy of Prof. G. Westbrook, University of Birmingham, UK.

Although the relatively fast-spreading Chile Rise does not have as extreme topography as the slower-spreading Mid-Atlantic Ridge, the effect on the trench system is quite dramatic. The plate motion is such that the oceanic plate is moving east beneath the continental crust. Because the Chile Trench trends northeast-southwest, the intersection point, known as a trench-trench-ridge triple junction, migrates northwards. The effect of this relative motion of the triple junction allows the effect of the ridge subduction to be seen, both before and after the collision.

To the north of the triple junction, the Chile Trench is characterised by a wide and deformed accretionary complex (Thornburg et al., 1990). However, this is much narrower and less deformed to the south of the triple junction. Also apparent from the GLORIA imagery of this region are the faults that are most prominent adjacent to, and to the south of, the triple junction. These faults, appearing as bright and dark reflectors on the GLORIA image, are approximately ridge-parallel. That they are most significant at and to the south of the ridge-trench collision is probably a result of their distance from the trench axis. Bending of the down-going oceanic plate oceanward of the trench reactivates ridge-parallel faults, formed during normal spreading. This spreading fabric forms an inherent weakness in the crust which is exploited by the bending forces acting on the plate as it approaches the outer trench wall. The GLORIA image shows these faults as facing towards the trench (i.e. linear reflectors to the west of the sonar and shadows to the east) which is consistent with down-faulting towards the trench axis.

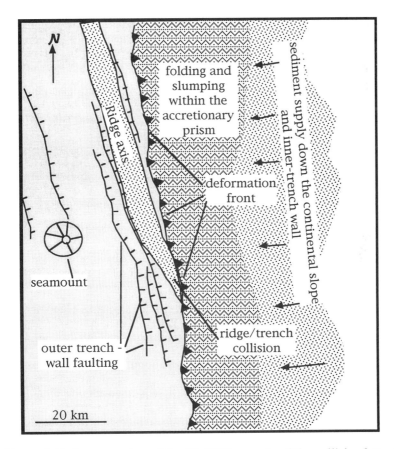

Figure 4.13. Interpretation of the GLORIA mosaic of the collision between the Chile Rise and the Chile Trench. The major features are shown and labelled: the barbed curvilinear line is the deformation front (barbs on the overthrust side), the lines with ticks are faults (ticks indicate the scarp facing directions), The chevron pattern denotes the area of deformation within the accretionary prism. The active spreading ridge is shown by the light stipple.

A major effect of the subduction of the Chile Rise into the Chile Trench, interpreted from the GLORIA data, is the erosion and subduction of the accretionary prism that lies between the ridge axis and the trench. The ridge acts like a bulldozer, deforming and compressing the accretionary prism in front of it as it collides with the trench. As the ridge is subducted, so it takes with it a large amount of the accretionary sedimentary complex (Cande and Leslie, 1986). Similar effects occur where large seamounts are subducted into trenches, creating arcuate deformation fronts, folds, slumps and slides.

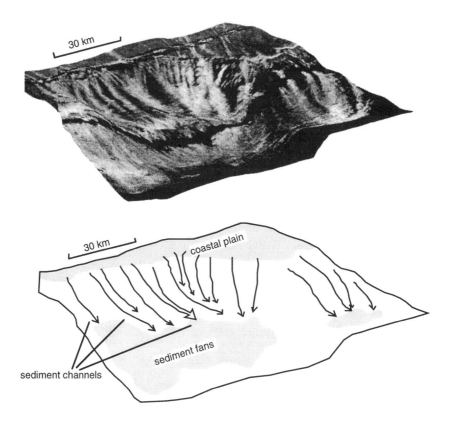

Figure 4.14. GLORIA image draped over bathymetry and projected as a perspective view plus line interpretation. The data are from the insular slope of Puerto Rico and reveal a cusp-shaped, slump scarp (Schwab et al., 1991). Copyright Elsevier Science Publishers BV, Amsterdam.

Another significant effect of the steep slopes found on the margins of deep-ocean trenches is the periodic collapse of the sediment pile on the inner trench wall. This is a result of two processes which are frequent at collision margins: the accumulation of poorly consolidated material on steepening submarine slopes and earthquake shocks. In the example, and accompanying line-drawing interpretation (Figure 4.14), GLORIA imagery is draped over bathymetry to reveal the relationship between bright branching reflectors (Schwab et al., 1991). This example comes from the Puerto Rico Trench, a relatively minor subduction system in the Caribbean that results from the convergence of the Caribbean and North American plates.

The perspective view shows the relationship between the sediment supply (channels and slumps) with the general shape and orientation of the cusp or scar. Cusps are formed by the mass removal of sediment on steep slopes by slumping. In this case, the cusp indicates the removal of some 1,500 km^3 of sediment which has slumped away from the insular margin into the Puerto Rico Trench. The cause of this is probably an increase in the angle of the slope by the continuing convergence of the plates (Schwab et al., 1991). Bright branching reflectors indicate coarser sediment channels that feed downslope from the coastal shelf, over the cusp and into the trench. As with the example of channels (e.g. Figure 4.10, the Columbian Trench) the coarser material appears to be derived from the shallower areas and flows downslope. Minor tributaries coalesce to form larger ones that ultimately feed sediment fans at the base of the slope. These fans appear as brighter regions near the lowermost region of the cusp. Darker, smooth regions on the image are softer and finer-grained areas of sediment. The chaotic bands of reflectors that run from left to right across the centre and upper parts of the image are from the nadir region of the GLORIA sidescan sonar and comprises noisy and poor data that are not geological features. This image is a good example of the combination of both bathymetry and sidescan sonar imagery.

4.6 MUD VOLCANOES AND SERPENTINITE SEAMOUNTS

The rapid accumulation of sediments forming the accretionary prism results in a build-up of high fluid pressures within the sediment pile. These sediments are a mixture of pelagic muds with a high water content and sands which have less water but are more permeable. This mixture of materials allows an over-pressurisation of the sediment-water mixture (Bangs et al., 1990). When an opportunity arises for the pressure to be released, by opening up a fluid pathway to the surface, the more fluid mud-water mixture moves upwards and flows out on to the seafloor (Henry, 1996). The result of this process is the formation of "mud-volcanoes" which range from a few tens of metres in diameter to several hundreds of metres (Brown and Westbrook, 1988). Examples of mud volcanoes are shown in Chapter 7 where the process forming them is similar, but in a passive margin environment.

The examples of mud volcanoes shown in Figure 4.15 are images taken by GLORIA sidescan sonar from the northern Columbian Trench. The ensonification direction is from the bottom left corner towards the top right corner. The features are identified as near-circular objects with a bright side towards the sonar and shadow cast on the far side. They range in diameter from a few hundred metres to a few kilometres, although they are only a few tens of metres high. It is interesting to note that the shape of the mud volcanoes becomes more elongate with distance away from the sonar. Whereas this has been interpreted to be a result of increasing deformation adjacent to a strike-slip fault zone indicated by the bright lineated region (Westbrook et al., 1995), a possible alternative explanation is far-range along-track blurring. This phenomenon, particularly marked for long-range low-frequency sidescan sonars such as GLORIA, results from beam spreading at increasing range (see Chapter 9). The far-range targets are thus ensonified by more pings from the sonar than targets of similar size in the near range. This repeated ensonification results in an apparent elongation of features in the far range and is responsible for the generally striped appearance of the image at a long distance from the instrument.

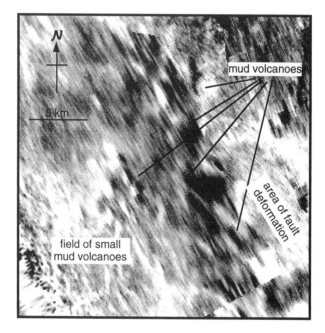

Figure 4.15. GLORIA imagery of mud volcanoes, both individual and a field of many, taken from the northern Columbian Trench (courtesy of Prof. G. Westbrook, University of Birmingham, UK). The volcanoes appear as circular objects with bright side towards the sonar and shadow on the far side. The increasing elongation of the objects along with an increasingly striped appearance of the image in the far range is probably an artefact of beam spreading. This is common for low-frequency, long-range sonars like GLORIA. Care must be taken not to confuse this effect with real geological processes.

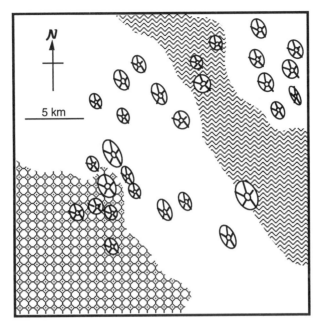

Figure 4.16. Line-drawing interpretation of the GLORIA imagery of mud volcanoes in the northern Columbian Trench. The volcanoes appear as circular objects although with increasing elongation at greater range from the sonar. This elongation may be real, or is more likely a result of beam spreading in the far range. The area in the lower left-hand corner of the scene is a field of smaller mud volcanoes. The obliquely trending wavy patterned area in the top right is a zone of deformation associated with a major strike-slip fault region.

The subduction of old oceanic crust back into the Earth's interior releases water at great depths. This process, known as dehydration, expels water upwards into the overlying mantle which is predominately formed from a mineral called olivine. At low temperatures, below 400°C, the water and olivine react to form a secondary mineral called serpentine. This mineral product has unusual properties; it is much less dense than the ocean crust and it tends to flow easily.

In the Mariana forearc, a number of large seamounts have been discovered that are not of volcanic origin. These have been investigated by sidescan sonar, multibeam bathymetry, submersibles and deep-ocean drilling. The conclusions of such studies have been that these features are volcanoes formed by the eruption of serpentinite on the seafloor (Fryer and Fryer, 1977). An analogy has been drawn between both mud volcanoes and hot molten lava volcanoes: the serpentinite volcanoes are formed by the cold eruption of a serpentine-water mixture which builds up an edifice from the summit of which serpentinite flows. However, unlike the mud volcanoes, and more analogous to conventional volcanic activity, the serpentinite is a direct product of the mantle lying beneath the ocean crust upon which the serpentinite volcanoes are formed.

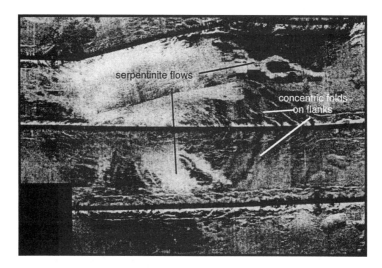

Figure 4.17. Example of SeaMARC II sidescan sonar imagery of a serpentinite seamount called "Conical Seamount" in the Mariana forearc (Fryer and Pearce, 1992). Copyright Ocean Drilling Programme, College Station, Texas, USA.

Figure 4.17 shows an example of SeaMARC II sidescan sonar imagery of a serpentinite seamount called "Conical Seamount" located in the Mariana forearc (Fryer and Pearce, 1992). These seamounts are similar to mud volcanoes in that they are density-driven eruptions of cold, mobile and poorly consolidated material. The bright reflective areas are serpentinite flows. Wavy tongues of serpentinite are clearly distinguishable, elongated towards the lower right-hand side of the image, where they have flowed downslope. Note in the top right-hand corner of the image, the lower reflective flow overlain by the brighter flow to the right of the summit (Figures 4.17 and 4.18). The decrease in acoustic reflectivity is probably caused by progressive burial of the flow by soft pelagic sediment that also blankets the remainder of the seafloor. The serpentinite flows have originated at the summit of the volcano (centre left).

Drilling of the serpentinite flows by the Ocean Drilling Programme have shown that they comprise a fine-grained serpentinite mud matrix that hosts blocks of solid rock that range from a few centimetres to several metres in size (Fryer and Pearce, 1992). The roughness of this serpentinite mud and rock mixture is probably responsible for the increased strength of backscatter from the lava outcrop on the seafloor. Concentric rings, especially visible as light and dark arcuate bands to the lower right of the summit, are undulations on the flanks of the serpentinite volcano.

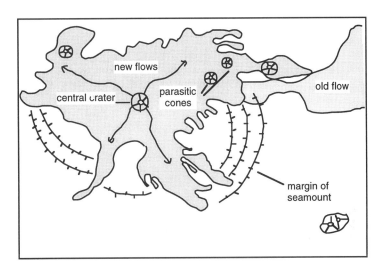

Figure 4.18. Line-drawing interpretation of the SeaMARC II sidescan sonar image of the Mariana forearc serpentinite seamount. The light stipple indicates the position of the brighter backscattering serpentinite flows. An older flow lies beneath the younger ones to the right of the image. The tongues of serpentinite are formed of a mixture of serpentinite mud and blocks of solid serpentinite. The concentric rings are probably a result of lateral flow of the flanks of the seamount, downslope, and away from the summit.

4.7 SUMMARY

Deep-ocean trenches, formed at collision margins, pose great technical challenges to their investigation. As the deepest places on the Earth's surface, the intense pressures at the bottom of these trenches preclude any direct observations. Sonar imagery is the only means to map the seafloor in these trenches, yet the great depths require high-powered systems that can give reliable data. Few, if any, deep-towed sidescan sonar systems have been deployed in the deep-ocean trenches. Most of the data are from surface-towed, low-frequency systems such as GLORIA. As a result, the resolution of the data are crude, limited to about 100 x 100 m. Despite this, the quality of the information derived from such sonars is of immense value. In recent years, the geological processes operative at these active plate margins has been revealed by sonar imagery: the accumulation of sediment piles to form accretionary complexes, slides and channels of sediment feeding the trenches, deformation of trench floors, and eruption of mud-volcanoes and serpentinite seamounts. As the loci of some of the most intense earthquakes on the planet, knowledge of the processes operating within the deep-ocean trenches is essential to human activity. Whether this is offshore engineering or predicting and avoiding natural disasters, knowledge of the environment of deep-ocean trenches will be increasingly in demand.

In this chapter, we have introduced some of the aspects of sidescan sonar imagery interpretation from deep-ocean trench environments. The thinking behind the interpretations has been explained and we have presented some of the best examples of sidescan sonar imagery of these environments available. The interpretation check-list can be applied to any sidescan sonar images, and it forms the basis of the interpretations explained in the following chapters. The images presented here can be used as reference material to allow comparison and aid interpretation of other, unknown, sidescan sonar images from deep-ocean trench environments.

4.8 FURTHER READING

An, L.-J., C.G. Sammis; "Development of strike-slip faults: shear experiments in granular materials and clay using a new technique", Journal of Structural Geology, vol. 18, no. 8, p. 1061-1077, 1996

Brown, K., G. K. Westbrook; "Mud diapirism and subcretion in the Barbados Ridge Accretionary Complex: the role of fluids in accretionary processes", Tectonics, vol. 7, p. 613-640, 1988

Cande, S.C., R.B. Leslie, J.C. Parra, M. Hobart; "Interaction between the Chile Ridge and Chile Trench: geophysical and geothermal evidence", Journal of Geophysical Research, 92(B1), p. 495-520, 1987

Cande, S.C., R.B. Leslie; "Late Cenozoic tectonics of the southern Chile Trench", Journal of Geophysical Research, 91(B1), p. 471-496, 1986

Fitch, T.J.; "Plate convergence, transcurrent faults and internal deformation adjacent to southeast Asia and the western Pacific", Journal of Geophysical Research, vol. 77, p. 4432-4460, 1972

Fryer, P., H. Fryer; "Origins of non-volcanic seamounts in a forearc environment". In *Seamounts, Islands and Atolls.*, B Keating, P. Fryer and R Batiza (eds) American Geophysical Union, AGU Monograph Series, 43, p. 61-69, 1977

Henry, P.; "Fluid flow in and around a mud volcano field seaward of the Barbados accretionary wedge: results from Manon cruise", Journal of Geophysical Research, vol. 101, no. B9, p. 20297-20323, 1996

Leggett J. (ed); "Trench-forearc geology: sedimentation and tectonics on modern and ancient active plate margins", Geol. Soc. Spec. Pub., 576 pp., 1982

Moore, G.F., T. H. Shipley, P. L. Stoffa, D. E. Karig, A. Taira, S. Koramoto, H. Tokuyama, K. Suyehiro; "Structure of the Nankai Trough accretionary zone from multichannel seismic reflection data", Journal of Geophysical research, vol. 95, p. 8735-8765, 1990.

Pickering, K.T., M.B. Underwood, A. Taira, J. Ashi; "IZANAGI sidescan sonar and high-resolution multichannel seismic reflection line interpretation of accretionary prism and trench, offshore Japan", in "*Atlas of Deep Water Environments: Architectural style in turbidite systems*". K. T. Pickering, R. N. Hiscott, N. H. Kenyon, F. R. Lucchi and R. D. A. Smith (eds), Chapman & Hall: London, p. 34-49, 1995

Schwab,W. C., W. W. Danforth, K. M. Scaanlon, D. G. Masson; "A giant submarine slope failure on the northern insular slope of Puerto Rico", Marine Geology, vol. 96, p. 237-246, 1991

Thornburg, T.M., L.D. Kulm, D.M. Hussong; "Submarine-fan development in the southern Chile Trench: a dynamic interplay of tectonics and sedimentation", Geological Society of America Bulletin, vol. 102, no. 12, p. 1658-1680, 1990.

Westbrook, G.K., M.J. Smith; "Long decollements and mud volcanoes; evidence from the Barbados Ridge complex for the role of high pore fluid pressure in the development of an accretionary complex", Geology, vol. 11, p. 279-283, 1983

Westbrook, G.K., N. Hardy, R. Heath; "Structure and tectonics of the Panama-Nazca plate boundary". Geol. Soc. Am. Spec. Paper 295, 1995

Woodcock, N.H., M. Fischer; "Strike-slip duplexes", Journal of Structural Geology, vol. 7, p. 725-735, 1986

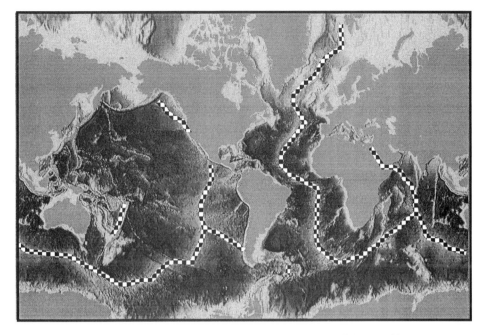

Figure 5.1. Location of mid-oceanic ridges around the world.

5

Mid-Ocean Ridge Environments

5.1 GEOLOGICAL BACKGROUND

Mid-ocean ridges are the loci of construction of the Earth's oceanic crust (Vine and Matthews, 1963). They are the most energetic geological environments on the planet. Characterised by high topographic relief, the mid-ocean ridges are dominated by volcanic and tectonic processes. Globally, over 60,000 km of mid-oceanic ridge produce ~35 km^3 of volcanic crust every year (Figure 5.1). This new crust is welded to the retreating edges of the older crust as the Earth's plates move apart resulting in chains of volcanoes, lava fields, fissures, faults and cliff scarps. The Mid-Atlantic Ridge is longer than the Himalayas and assumes the form of a rift. It is a volcanic chain split along its entire length by a deep axial valley. On the other side of the Earth, the East Pacific Rise has the appearance of a smooth line of hills or a gently undulating ridge.

Along the entire length of all mid-ocean ridges, volcanoes of every shape and size dominate. Offsets in the trend of the mid-oceanic ridges disrupt the continuation of the volcanic chain. These fracture zones are deep, linear trenches that stretch from one side of the ocean basins to the other.

Faults and fissures, caused by the forces stretching the ridge and pulling the global tectonic plates apart, cut into the newly formed crust. On slow-spreading ridges these faults can displace the crust by several kilometres exposing the underlying mantle. At faster-spreading ridges, the faults are less dramatic, but both step down and up away from the rift axis. In all mid-ocean ridges, faults play a dominant part in controlling the location of volcanism and hydrothermal activity.

With lava approaching temperatures of 1,200°C, the thermal energy released by volcanic eruptions along the mid-ocean ridges is of the order of five million gigawatts. This is many times the amount of power generated by the total combined activity of human civilisation today. Of this energy, about 1% is transferred through high-temperature hydrothermal vents, jets of superheated water that gush out of the ocean floor.

Hydrothermal vents, called black smokers, discharge mineral-rich waters heated by the underlying molten rock and expelled through cracks in the young ocean crust (Rona, 1986). The water temperature within the black smokers, which approaches 360°C, carries many dissolved metal species (iron, manganese, copper, etc.). Each black smoker releases tonnes of these metals into the ocean each year. There are probably thousands of black smokers on the global mid-ocean ridge system, which together release enormous amounts of metals into the environment every year. A part of these metals collects on the seafloor close to the vents and as a result builds up valuable mineral deposits forming mounds up to several tens of metres high and hundreds of metres in diameter. Apart from forming valuable mineral reserves, black smokers support exotic life forms that have evolved outside the influence of sunlight. High-temperature bacteria, that colonise the mineral deposits, are considered by some microbiologists to be valuable resources for biotechnological exploitation (e.g. Tunnicliffe, 1991).

All geological aspects of these environments are visible with sidescan sonar. The degree to which details can be imaged depends on many factors: the type and frequency of sonar, the size of area and speed at which a survey is required; all these requirements determine the scale at which imagery is to be gathered. Examples of sidescan sonar images of mid-ocean ridge environments, and their interpretations, are given in the following sections. Recent developments in deep-towed sidescan sonars have revolutionised our view of mid-ocean ridges. Therefore, many of the examples come from such sonar systems.

5.2 SHAPE DERIVATION FROM BACKSCATTER

One of the ultimate objectives when interpreting sidescan sonar images is to understand the shapes of objects. In areas such as the mid-ocean ridges the topographic relief is responsible for the majority of the backscattered sonar signals. Therefore, sidescan sonar images of such terrains reveal information primarily about the local slopes and shapes of objects within a scene.

To derive an impression of the shape of objects requires an understanding of the direction of ensonification, the grazing angles of the incident beams and details about the grey scale values of the pixels. The interpretative principle depends on the fact that acoustic backscatter varies with increasing incidence angle (see Chapter 2). If we assume that the backscatter values are largely a result of incidence angle, then the high values correspond to steep slopes facing the vehicle and the low values to flatter areas. The zero backscatter values, i.e. the shadows, correspond to slopes facing away from the vehicle that are steeper than the angle of the incoming sonar beam. Thus the histogram of backscatter values, along a profile perpendicular to the sonar, can be considered as a derivative of the seafloor slope. To derive a shape profile, first the backscatter values must be corrected for the value that most closely corresponds to flat seafloor. As an initial approximation, this value can be an average of the scene. This value is then removed from the backscatter values such that shadow values become negative. The cumulative histogram of backscatter values, summed in the direction of ensonification, performs an integration and gives a relative shape profile of the object.

The DSL-120 image of the TAG hydrothermal mound (Figure 5.2, upper) is taken as an example of how to derive the object's shape from the acoustic backscatter represented by the pixel values. The image has been acquired from a deep-tow vehicle passing above and to the left-hand side of the scene. Bright pixels represent high backscatter values. Black pixels are shadows and have grey levels with a zero value. The scene is approximately 175 m square.

The variation of pixel values (samples) with distance from the vehicle along line 87 (shown) is displayed in the histogram shown below (Figure 5.3, lower panel). Backscatter values range from 0 to 255. The correlation between the position of bright pixels on the image and the high pixel values on the histogram can be seen by comparing the histogram with the line drawn on the image.

Initial inspection of the DSL-120 image reveals a series of concentric rings. The sides of the rings facing the vehicle have high backscatter values and are therefore slopes facing to the left. The opposite sides of rings have low backscatter values, caused by slopes facing away from the vehicle and casting shadows. From this it is possible to say that the object comprises several nested circular mounds and moats. This interpretation is confirmed by high resolution bathymetry for the TAG mound (Humphris et al., 1995) shown on Plate 1 (in the colour section of the book).

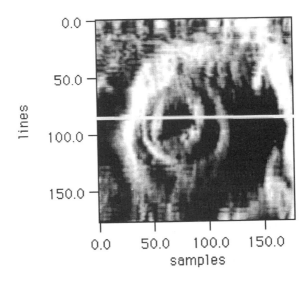

Figure 5.2. A sidescan sonar image of the TAG (Trans-Atlantic Geotraverse) hydrothermal mound (Humphris et al., 1995) made with the DSL-120 deep-towed vehicle. The image, ensonified from the left, is about 175 m square.

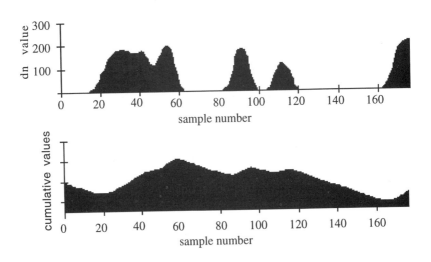

Figure 5.3. (top) histogram of pixel values with distance across line 87. Pixel values are proportional to the local slope facing the sonar; (bottom) histogram of cumulative pixel values with distance, corrected for the average value within the scene. Although there is no vertical scale, this type of shape profile gives an impression of the shape of the object.

From the shape profile, the local topography of the mound becomes immediately clear (Figure 5.3). By comparison between the slope profile and line 87 on the image, the peaks and troughs are seen to coincide with the positions of the concentric ring structure. The local high intersecting line 87 between sample positions 50 and 65 is clearly shown on the shape profile to be the highest point on the mound in the line of section.

The second example, shown in Figure 5.4, is from a 2 km wide, 30-kHz TOBI image of a series of left-facing fault scarps returning strong backscatter (bright pixels), and back-tilted blocks behind the scarps casting shadows (black pixels). The grey pixels are flatter areas of the seafloor and their average value has been used to adjust the cumulative histogram for the grazing angle of the sonar beam. Since the vehicle passed to the left of the scene, the backscatter values have been summed along a direction from left to right to derive the shape profile. From the image alone, it is unclear that the terrain shallows to the right, but this becomes apparent from the shape profile (Figure 5.4). As with the DSL-120 example of the TAG mound above, the derived shape profile is only a qualitative representation of the actual bathymetry, yet it is a valuable aid to interpreting the scene.

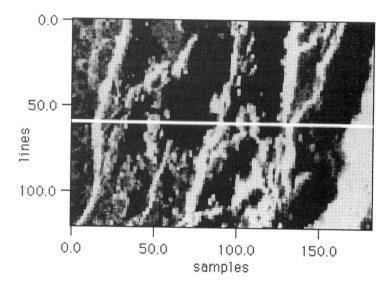

Figure 5.4. A sidescan sonar image of part of the western wall of the Mid-Atlantic Ridge (Murton, 1993) made with the 30-kHz TOBI deep-towed sonar. The image, ensonified from the left, is approximately 1,500 m long and 2,000 m wide.

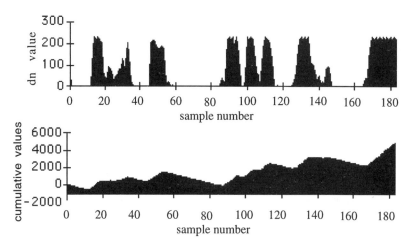

Figure 5.5. The upper histogram is a plot of pixel values with distance across line 60 on Figure 5.4. Again the value of the pixels approximates to the angle of the local slope. The lower histogram also shows the cumulative, corrected pixel values across line 60. The profile clearly shows the strong backscattered signals result from a series of steps (fault scarps) facing the vehicle, causing the terrain to shallow to the right.

It may be possible, with good quality geometrically and radiometrically corrected sidescan sonar data, knowledge of the vehicle height and a reliable backscattering law, to predict the bathymetry. However, in practice, noise and non-systematic variations in signal strength between successive pings causes adjacent shape profiles not to match and integrated errors to become large. Also, deviations from the slope-based reflectivity model, caused by textural heterogeneity and variation in acoustic facies, will give spurious slope results. However, with sophisticated processing, shape from backscatter can be used reliably to modify low-resolution bathymetry data by combining sidescan sonar with local depths derived from conventional soundings.

5.3 VOLCANIC FEATURES

Mid-ocean ridges are dominated by two fundamental processes: volcanic construction and tectonic dismemberment (Pezard et al., 1992; Murton and Parson, 1993). Fast (more than 5 cm per year) and hot (shallower than 2,500 m) spreading ridges are characterised by point-source seamount volcanoes and sheet flows (Gente et al., 1986). Tectonic activity on fast spreading ridges results in a dense population of small throw fault scarps that face both into and away from the ridge axis. Cold and slow ridges have largely composite volcanoes, hummocky volcanic ridges and pillow lavas. Tectonic activity here results in elevated topography dominated by steep and high inward-facing fault scarps bounding a central axial valley.

The following sections examine the features that are the basic units of construction of the oceanic crust: the lava flows, volcanic edifices, volcanoes, volcanic clusters and axial volcanic ridges. This is followed by an examination of features resulting from the processes that modify the oceanic crust: fissures, faults, scarps, and transform and non-transform discontinuities. Finally, regional sonar images of parts of the global mid-ocean ridge system are presented where the basic features are seen together in a broader context

5.3.1 Point-source volcanoes

Small point-source volcanoes, ranging from a few hundred metres to a kilometre in diameter and up to a few hundred metres high, are among the very basic units of construction of the oceanic crust (Smith and Cann, 1990). The sonograph of Figure 5.6 shows a small cratered conical volcano, typical of this type of feature. It is shown again, for comparison, in Figure 5.7 where it has been ensonified by another sonar system. In the following pages, a variety of images of individual volcanic edifices are presented along with diagrammatic interpretations and the methodology for their interpretation explained.

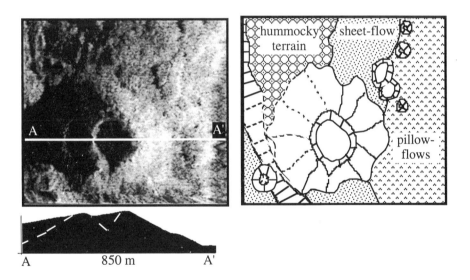

Figure 5.6. Left: TOBI image of a volcano on the Reykjanes Ridge, at a depth of 1,850 m. Right: line diagram illustrating the roughly circular volcanic cone, central crater and various surrounding terrains. The shape profile, in silhouette, is derived from an integration of pixel values along the line A-A'. Dashed lines represent the profile hidden by the shadow. This image can be compared with an image of the same feature made by the lower resolution SeaMARC II sonar (Figure 5.7).

The scene in Figure 5.6 contains a roughly circular black shape with a bright rim, with a bright crescent to the right and a dark one to the left. Surrounding this is a region of variable texture that is brighter to the right and darker to the left. The circular shape is a central summit crater. Its bright rim is the crater wall, with strong echoes returned from the inner wall on the left-hand side. The presence of a crater, combined with the smooth textured flanks, indicates that it is a monogenic eruptive volcano. Fine details on the image reveal a dark "V"-shaped region in the southern wall of the crater that is interpreted as a small cleft or breach in the rim.

Variable textured areas, appearing as small rounded features on the sonograph, indicate a number of individual, yet coalesced, flat-topped volcanic mounds to the north-east of the volcano and rougher hummocky terrain to the north-west. Brighter areas to the right and dark areas to the west of the image are reflective sheet flows and dark sediments respectively.

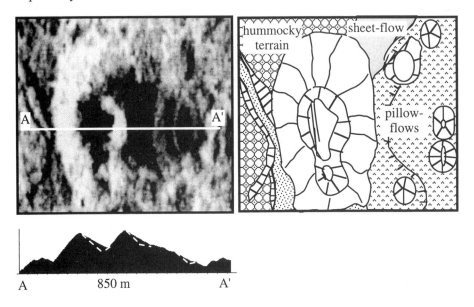

Figure 5.7. Here, the same monogenic volcano as the one illustrated in Figure 5.6 is shown, but imaged by the lower frequency, lower resolution, surface-towed sonar SeaMARC-II (courtesy S. Shor, University of Hawaii, USA). Differences in the resolution and acquisition attitude, including different ensonification directions, between the SeaMARC II and TOBI sonar systems can be seen by comparing the two images. Despite this, the shape profile from along the line A-A' compares well with the same profile on Figure 5.6.

Flanking the volcano, the side facing the TOBI vehicle is characterised by a bright speckled texture. In contrast, the surrounding terrain has a lower degree of contrast variation. These differences indicate scree on the flanks of the volcanic edifice and a fine-grained sediment drape on the surrounding seafloor. Scree is composed of angular rock fragments, which have a large size ranging from several centimetres up to a metre, and which are highly unsorted. In this environment, at a depth of 3,200 m and far from the continental shelf, the fine-grained sediment is likely to be of pelagic origin, which is a combination of mud and carbonate ooze.

The shape profile in Figure 5.6, derived from the TOBI image along line A-A', is dominated by the flank of the volcano facing the sonar. Because the sonar vehicle was at a low altitude compared with the height of the volcano, the crater and far side of the feature are dominated by shadow. Both the crater and far side flank have slopes greater than the shadow angle, hence the interpreted profile is shown by the dashed line.

Figure 5.7 shows a SeaMARC II sonograph of the same seamount on the crest of the Reykjanes Ridge as shown in Figure 5.6. Despite differences between the sonars, the seamount looks remarkably similar in both images. The higher altitude of the surface-towed SeaMARC II has allowed a fuller ensonification of the crater interior and the slightly elongate shape of the volcano, stretched along track, may be a result of different anamorphic corrections applied to the images for the sonars' speed over the ground.

Although some of the flat-topped mounds to the east of the volcano can readily be identified from the SeaMARC II image, the general fuzziness of the SeaMARC II image is a result of a wider beam angle and hence greater beam width and a greater range to target. With such lower resolution images, it becomes more difficult to distinguish between subtly different seafloor textures such as variations in the degree of pelagic sediment drape. Compared with the TOBI sonograph, the SeaMARC II image less precisely distinguishes between the different terrains or separates the coalesced flat-topped volcanic mounds.

The SeaMARC II derived shape profile reveals the depth of the central crater more accurately than the TOBI derived shape profile. This is because the higher angle of incidence of the incoming SeaMARC II sonar beam reduces the shadow length, resulting in more information from slopes facing away from the vehicle.

A major difference between the TOBI and SeaMARC II images has been the direction from which the sonar ensonified the scene. Comparison between the images reveals details in one that are hidden by shadow in the other. For example, eastward-facing faults to the west of the volcano are apparent only on the SeaMARC II image. An arcuate reflector appearing within the shadow region on the west flank of the volcano is only apparent on the TOBI image suggesting that it is an artefact. This reflector has the same shape as the western rim of the summit crater. Yet it is displaced west of the rim by a distance equivalent to the width of the crater. It is possible that this reflector is in fact a multiple caused by an internal echo from within the crater itself.

These differences between the images of the same feature do not indicate that one system is better than another. Rather, the different acquisition parameters of the two sonars have both advantages and disadvantages. For example, the SeaMARC II data were acquired in a fraction of the time compared with the TOBI data. Thus the SeaMARC II system allows relatively high-resolution imagery, with less shadow, at a high acquisition rate. TOBI allows higher-resolution imagery, offering greater discrimination of seafloor terrain types, yet with more shadow and at a significantly slower acquisition rate.

5.3.2 Composite volcanoes

A TOBI 30-kHz sonar image of a small volcanic cone from 27°35'N on the Mid-Atlantic Ridge is shown in Figure 5.8. The image reveals a characteristic circular shape and central summit crater, typical of this common type of conical volcano. The central, circular dark feature is a deep summit crater. The bright smooth crescent to the right-hand side of the crater is a scree-covered flank facing the sonar. The chequered appearance of the image is a function of its high magnification in which each pixel is recognisable. The mottled texture to the north and south of this volcano reveals multiple parasitic cones and hummocks (i.e. smaller volcanic cones and mounds), indicating a composite construction.

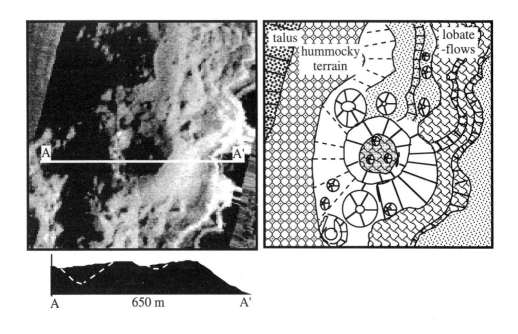

Figure 5.8. TOBI sonograph (top) of a small composite conical volcano on the Mid-Atlantic Ridge at 27°35'N and at a depth of 3,400 m. The line drawing (middle) and shape profile (lower) show the various regions comprising the volcano and the shape of the feature respectively.

Composite volcanoes are commonly formed by slow and episodic lava eruptions. Despite this, that the edifice conforms to a single conical shape with a central crater is evidence that the main conduit for magma egress to the surface was mainly through a single point that was stable during the eruption of the main volume of the volcano. The parasitic cones and small volcanic piles probably post-date the construction of the main cone. Central summit craters are common on all types of volcanoes on mid-ocean ridges and suggest subsidence of magma in the conduit or collapse of a high-level magma body close to the base of the volcano. Parasitic cones on the floor of the crater, the tops of which are just visible on this image, are late-stage eruptions after the crater subsided.

Curvilinear, bright areas facing to the right, and dominating the right-hand flank of the volcano, are lava flow fronts. The smooth textured area on the northern flank and summit ridge of the volcano may be either sediment ponds or smooth sheet flows. The latter is the more probable explanation in this example for two reasons: the rest of the scene has a high-contrast variation indicative of sparse sediment cover on rougher volcanic material, and the lobate flow fronts suggest relatively high effusion rates in the volcano's history of which the occurrence of sheet flows would be consistent.

The speckled area to the far left-hand side of the scene, against which the volcano's shadow is thrown, is interpreted as scree formed on the slope of a steep fault scarp that faces towards the sonar vehicle. The shape profile (A-A') is dominated by the concave slope of the volcano flank. Shadows obscure the crater and far flank slopes and have been interpreted (dashed lines).

5.3.3 Central volcanoes

A major volcanic unit on spreading ridges is the central volcano. These volcanic systems are characterised by a large point-source volcanic edifice with flanking ridges. They range in diameter from several hundreds of metres to several kilometres. One of the best examples comes from the Mid-Atlantic Ridge at 28°55'N and is known as W-seamount (Cann and Walker, 1993). In Figure 5.9, the W-seamount has been imaged by TOBI (ensonified from the bottom to the top of the scene) at an altitude of 400 m. It is characterised by: a deep and well-defined summit crater (the circular shadow in the centre of the scene), with a flat rim (the slightly mottled circle of mid-grey tones surrounding the central crater) that has breaches to its left and right sides (dark lines, trending left-right, cutting the rim). A bright crescent on the far side of the crater is the inner crater wall.

Flanking ridges, also to the left and right of the volcano, are typical of a central edifice controlled by an underlying fissure system. On top of the flanking ridges are a number of rounded shadows with adjacent crescent-shaped bright areas in front. These are smaller volcanic cones and hummocks, a few hundred metres across, that have coalesced to form the ridges flanking the central edifice. The volcano's flank facing the sonar has a speckled and streaked appearance, with streaks forming a triangle-shaped brighter area (bottom centre of the scene). This is interpreted as a scree-fan with stone chutes running down the flank of the volcano. Behind this central volcano the terrain is characterised by a mottled texture that indicates a hummocky volcanic field.

In the diagrammatic interpretation of the W-seamount image (Figure 5.10), a characteristic feature is the flat-topped crater rim that dominates the shape of the volcano. Flat tops to submarine volcanoes have been interpreted as an indication that the edifice has achieved its maximum height in relation to the depth of its underlying feeder magma chamber (Cann et al., 1992). Once this has happened, further lavas that reach the summit flow horizontally, increasing the volcano's diameter and forming a flat top.

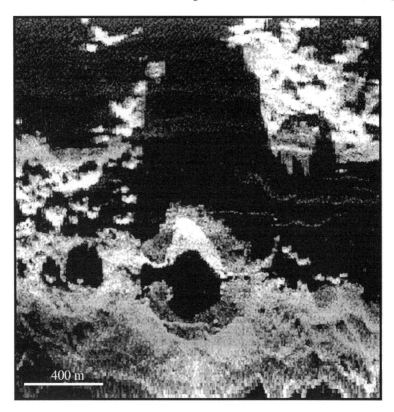

Figure 5.9. W-seamount, at 28°55'N on the Mid-Atlantic Ridge, ensonified by TOBI (from the bottom to the top of the scene) at an altitude of 400 m above the seafloor. The dominant features are the flat-topped crater rim, deep summit crater and left-right oriented flanking ridges that dominate the shape of this central volcano system.

The deep summit crater suggests formation by lava lake deflation and its breached crater rim by lava overspill and flow along the flanking ridges to the left and right of the summit. The large amount of scree on the flanks of the volcano, as opposed to fresh lava flows, indicates both a steep slope and a relatively old age for the edifice. This interpretation is supported by the accumulation of sediments both in front of and behind the volcano.

Yet despite the evidence for an old relative age, the lack of faulting of the edifice indicates that it has not experienced as much tectonic deformation as may be expected for a feature of this kind located in the axial valley of the Mid-Atlantic Ridge. Parasitic cones (smaller late-stage volcanic piles) to the left and right along the flanking ridges, and a few left-right oriented faults and grabens across the volcano, attest to an underlying fissure control on the location of the feature. Fluid dynamic models of fissure eruptions and evidence from Iceland of similar phenomena suggest that vertical

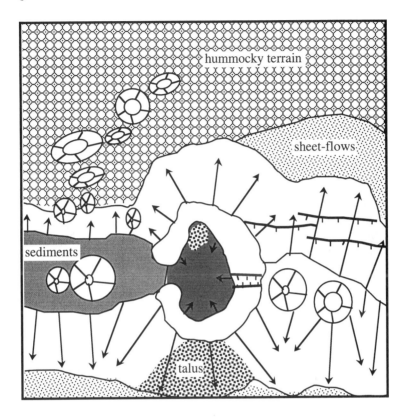

Figure 5.10. Line drawing of the W-seamount image shown on Figure 5.9 and at the same scale. Arrows point downslope. Note the deep central crater, flat rim and scree on the flank and the inner rim wall facing the sonar.

sheet eruptions become unstable and readily become point-source eruptions centred on the middle of the fissure system that supplies the magma to the surface. Where such systems evolve from fissure to point-source eruptions, central volcanoes result.

When a bathymetric model of the seafloor is available, for example from multibeam swath sonar systems, and is co-registered with the sidescan sonar imagery then perspective views can be generated. In the case of the TOBI image of the W-seamount, multibeam swath bathymetry was acquired separately from the sidescan sonar data and later co-registered within a common geographical reference frame. The coarser horizontal resolution of the swath bathymetry grid (typically with 150 x 150 m pixels) must be interpolated to have the same resolution as that of the sidescan sonar image (for TOBI, typically a pixel size of 6 x 6 m). The sidescan sonar image is then mapped texturally on to the interpolated swath bathymetry grid and a perspective view projected from a given azimuth and elevation.

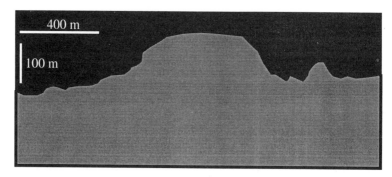

Figure 5.11. Skyline profile of the W-seamount (2 x vertical exaggeration), derived from scaling the shadow cast by the feature on the seafloor. The elevation of the volcano above the horizon (the scaling factor applied to the shadow) is a function of the sonar's range and altitude (see later sections). It is interesting to compare this skyline profile with the actual perspective view generated from co-registered bathymetry with the sidescan sonar imagery (Figure 5.12).

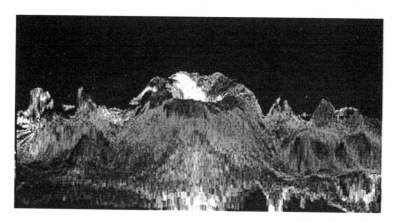

Figure 5.12. The sidescan sonar image (Figure 5.10) has been co-registered with, and draped over, swath bathymetry data for the same feature. A perspective view has then been projected depicting the feature as if the viewer were standing some 2 km away at an altitude of 300 m above the seafloor. The image has a vertical exaggeration of 2 and a horizontal scale identical to that of Figure 5.10.

Apart from giving a striking picture of the seafloor, this method of texture mapping sidescan sonar images on to bathymetric models has many geological applications. For example, sidescan sonar echo intensities are a function of both the incidence angle between the incoming sonar beam and the slope of the seafloor, and the seafloor roughness. Therefore a comparison between the shape of a feature and the location of strong echoes can help discriminate between seafloor slope and roughness. In Figure 5.12, returns from the far inside wall of the summit crater are stronger than may be expected from the slope alone. Thus we can conclude that the far inside wall of the crater has a rougher texture than for similarly oriented slopes elsewhere on the volcano. Similarly the conical shape of the talus scree fan, with its apex near the centre of the summit of the volcano, is more apparent in this perspective view because of its rougher texture returning stronger backscatter.

A further aid to the interpretation of co-registered sidescan sonar images with bathymetry grids comes from the use of depth colour coding (see Chapter 10 for a fuller explanation of the technique). This is illustrated on Plate 2 (colour section of the book). The colour is a function of the depth of the seafloor and the intensity a function of the strength of the backscattered sonar energy. When a perspective projection is made of the scene, the colour again helps the viewer to orientate the features of the image that have similar depths.

5.3.4 Flat-topped volcanoes

Flat-topped volcanoes occur in regions where the effusion rate of lava eruption is relatively fast and where the volcanoes have achieved their maximum height (Magde and Smith, 1995). The TOBI image of the flat-topped volcano, shown on Figure 5.13, comprises five discrete terrains: the roughly circular volcano and its round crater, small circular sedimented volcanic cones towards the top of the scene, surrounding sedimented hummocky volcanics, talus and scree in the upper right-hand side, and fresher hummocky volcanics towards the lower left-hand side of the scene. These terrains are illustrated on the line diagram interpretation (Figure 5.15). The flanks of the volcano are difficult to identify because they are steep. Also lay-over (displacement towards the sonar of images of elevated features) has distorted the image by pulling the image of the flank closest to the sonar in towards the vehicle's track.

The smooth texture and low backscatter of the top of the volcano are interpreted as relatively thick pelagic sediment cover. This interpretation is supported by the poorly defined summit crater, that is almost obscured by the sediment drape. Likewise, the low backscatter, locally smooth yet mottled texture surrounding the edifice is again interpreted as sediment-draped hummocky volcanic terrain.

The uneven shape of the shadow on the seafloor behind the volcano indicates that this area is uneven and probably varies significantly in elevation compared with the height of the volcano. From the shape profile, it is shown that the area immediately to the left of the volcano is probably deeper than the area to the right of it.

In the far left-hand side of the image, individual hummocks are identifiable that, together with their brighter backscatter intensities, suggests that they are fresher areas of hummocky volcanics. Other mounds are seen on the lower left-hand side of the flat-topped volcano, and on close inspection one at least has a central shadow indicating a small crater. These are probably parasitic cones, formed at a late stage in the evolution of the volcano.

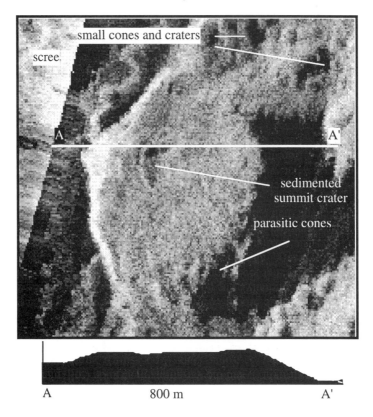

Figure 5.13. TOBI 30-kHz sonograph of a flat-topped seamount on the Mid-Atlantic Ridge at 28°35'N and a depth of 3,200 m. Distortions of the imaged edifice are caused by its proximity to the nadir of the sonar. The shape profile A-A' reveals the table-top shape of the volcano, its shallow central crater and a steep right-hand flank.

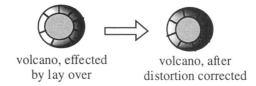

Figure 5.14. Lay-over is a distortion of elevated objects caused by slant-range correction solutions that assume a flat seafloor (see Chapter 10). The effect of lay-over is to pull the image of the highest part of an object towards the sonar. This effect increases with the height of the object being imaged and its proximity to the sonar.

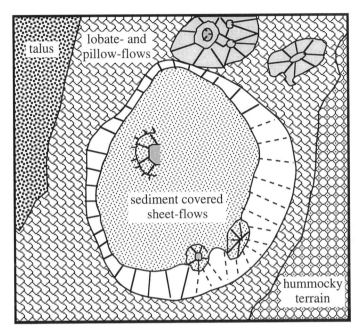

Figure 5.15. Interpretive line-drawing of the flat-topped volcano imaged by
TOBI and presented in Figure 5.13. The scene is approximately 1 km wide.

Near-circular features, larger than the hummocks yet smaller than the volcano, are
distinguishable in the upper centre of the scene. These smooth, locally homogeneous,
textured features are interpreted as small volcanic edifices. One of these, just north of
the main volcano, has a shadow in its centre indicating a summit crater.

The shape profile, derived from the TOBI image, shows steep-sided flanks and a nearly
flat top for this type of volcano. To the left of the centre of the volcano, the dark semi-
circular feature that also corresponds to a local hollow on the shape profile, indicates a
small localised summit crater. Such craters are often the result of subsidence of high-
level crustal magma chambers.

Flat-topped volcanoes, such as the one imaged by TOBI above, are generally thought to
be monogenic features, and are common on mid-ocean ridges. Unlike the conical or
hummocky volcanoes, the smooth texture and low ratio of their top to bottom diameters
suggest formation by either low-viscosity lavas or high effusion rates. The viscosity of
basaltic lavas is dominated by their silica content. However, on mid-ocean ridges the
silica content of basaltic lavas, which are the most common rock type, is restricted to
around 48%. Therefore, the control on the shape of lava flows and the edifice they form
is largely a function of the effusion rate. Given this, then flat-topped volcanoes are
thought to reflect high effusion rates during their formation. Therefore we would expect
sheet flows to be found on and around flat-topped volcanoes such as the one shown
above. The flat top of the volcano is though to be a result of the volcano having reached

its maximum elevation. This is controlled by the magma pressure which in turn is a function of the depth of the magma reservoir underlying the edifice (Smith and Cann, 1992). Thus, we have further evidence to suggest that the tops of these volcanoes are capped by horizontal sheet flows.

The smooth texture of the flat top of the volcano contrasts with the undulating tones of the surrounding floor. Since both terrains are almost certainly of comparable age, they are likely to have the same amount of pelagic sediment cover. Hence the smooth texture of the top of the volcano indicates sediment covering a flat underlying basement of sheet lava flows and the undulating terrain comprises sedimented hummocky pillow lavas.

5.3.5 Clustered volcanoes

It is quite common on spreading ridges for point-source volcanoes to be clustered together (Smith et al., 1995). Recent identification of abundant clusters of volcanic edifices on medium-spreading ridges has led some scientists to propose a new model for the formation of oceanic crust. This new model involves a "plum pudding" process in which the middle and lower oceanic crust comprise a multitude of discrete magma bodies of differing sizes and at different levels. Each magma body feeds a distinct volcano (Cann et al., 1992). Such clusters of volcanoes contain a range of types, from conical-cratered to flat-topped. Examples of volcano clusters are shown, with their interpretations, in the following figures.

In the images shown in Figures 5.16 and 5.18, a cluster of volcanoes on the Reykjanes Ridge, south-west of Iceland, are ensonified by both SeaMARC II and TOBI sidescan sonar systems. Below each image, the shape profiles, derived from the imagery along the lines A-A' in each image are shown respectively. The dominant features on both images are the circular-shaped objects that form the major part of the right-hand side of the images. From the disposition of the bright and dark crescent-shaped reflectors surrounding the circular objects, we can recognise slopes facing both towards and away from the sonars. Thus the circular objects are clearly topographic elevations. In a mid-oceanic ridge environment, which is dominated by volcanism, we can assume that these circular objects are volcanoes. The SeaMARC II images resolves three circular features forming the left half of the image, but the wider beam angle of this sonar leads to along-track elongation of targets that can be mistaken for track-parallel tectonic structures. This interpretation is easier to make from the TOBI imagery where circular features within the centres of the larger circles can be recognised as craters at the summits of the volcanoes.

The major difference between the two images is the resolution of different shapes, separation of features and the distinction of terrains. The higher altitude from which the SeaMARC II system ensonified the volcanoes, compared with the deep-towed TOBI system, has resulted in a more even illumination of the field. This is illustrated by the relative heights of the shape profiles with the one derived from the TOBI image being skewed to the right as a result of near-range foreshortening, otherwise known as lay-over. Differences between the shape profiles derived from the two images are largely a function of the lay-over caused by the closer proximity of the deep TOBI vehicle to the near-track features compared with the shallow SeaMARC-II.

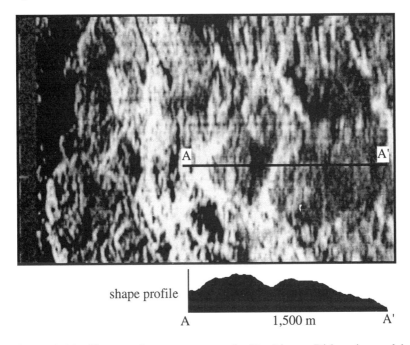

Figure 5.16. Cluster of seamounts on the Reykjanes Ridge, imaged by SeaMARC II (ensonification from left to right), courtesy of S. Shor, University of Hawaii, USA.

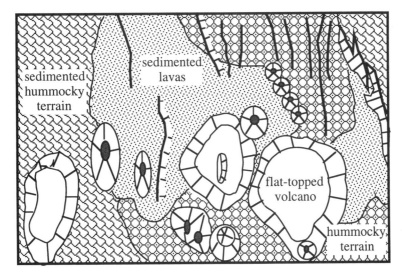

Figure 5.17. Interpretation of the SeaMARC II scene. There are essentially four different terrains: circular volcanoes, sedimented hummocky volcanics, fresh hummocky volcanics and sediment ponds.

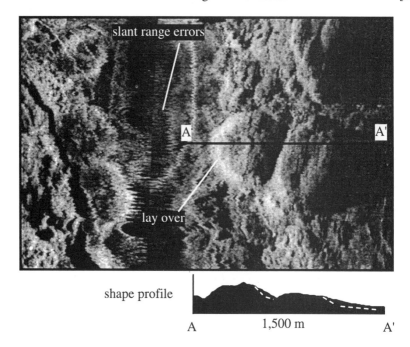

Figure 5.18. This TOBI image, of the same cluster of volcanoes on the Reykjanes Ridge as imaged by SeaMARC II in Figure 5.16, is subject to slant-range errors caused by incorrect determination of the vehicle's altitude.

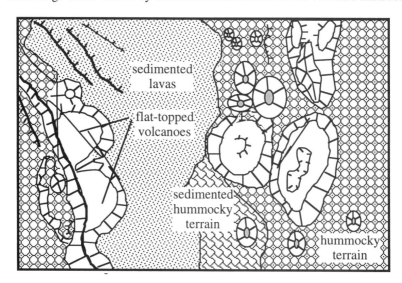

Figure 5.19. Interpretation of the TOBI image of the volcano cluster on the Reykjanes Ridge.

Compared with TOBI, SeaMARC II, being a surface-towed instrument, ensonified a greater amount of the field and has lost less of the image to shadows. However, its poorer resolution and wider beam angle mean the SeaMARC II image fails to differentiate many of the more subtle features, is unable to separate some of the larger structures, and is subject to track-parallel artefacts. The track-parallel features, in particular, are easily mistaken for tectonic lineaments and can give a false impression of the seafloor structure and age. Track-parallel artefacts are caused by the beam width, range to target and any yaw in the vehicle, to which surface-towed instruments are more prone than their deep-towed counterparts.

The TOBI image, acquired by the deep-towed ocean bottom vehicle, suffers from greater shadow areas and propagated slant-range errors derived from the nadir (appearing as zigzag across-track distortions). The zigzag distortions in the slant-range correction are caused by oscillation in the bottom echo detection. Although eventually corrected on the vehicle's hardware, the oscillation can also be overcome during pre-processing by using a running mean filter on the altitude file to smooth the slant-range variations.

The greater resolution of the TOBI image reveals considerable detail among and between the volcano cluster that is otherwise indistinguishable on the SeaMARC II image. Different components of the three major volcanoes are apparent such as: the rimmed central crater on the volcano closest to the start of the shape profile line (point A), composite summit mounds on the volcano immediately to the north-east and individual hummocks between the volcanoes. Likewise, the narrower beam width and shorter swath range of the TOBI system reduce far-range along-track blurring.

The different acquisition parameters of the two systems illustrated here have both advantages as well as disadvantages. The SeaMARC II system combines a high acquisition speed (typically 10 knots) with accurate positioning (effectively the ship's GPS position plus a few hundred metres of cable). In contrast the TOBI system acquires data at 1 to 2 knots, is relatively poorly positioned (ship's GPS position plus 5 to 10 km of cable) and requires a detailed chart of the seafloor bathymetry to allow safe deployment.

We can look at the TOBI image (Figure 5.20) of another volcano cluster and divide the terrain into three essentially different components. Being a mid-oceanic environment, we recognise the circular features as volcanoes and the rough, highly contrasting terrain as hummocky volcanic terrain. Smoother areas on the image, with low contrast variation, are interpreted as sedimented seafloor. Crossing the image are linear features, both as bright and dark lines, which correspond to faults. Depending on whether they appear as either bright or dark features, the faults face towards or away from the vehicle respectively.

Not all tectonic features are identifiable directly (see section 5.4 for further details). For example, the summit of the volcano that lies close to the centre of the scene but just to the upper left-hand side of the sonar vehicle track has a crater that is slightly offset. The offset is linear and along its strike a series of small shadows can be traced. We interpret this as indirect evidence for a small fault or fracture that is oriented north-south and which traverses the summit and crater of this volcano. A more detailed discussion about tectonic features, how to recognise them, and how to extract structural trends, is presented in a later section.

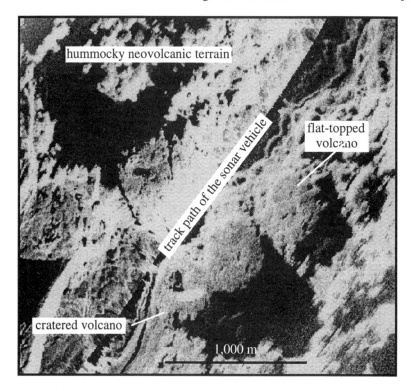

Figure 5.20. This TOBI image of a cluster of volcanoes on the Mid-Atlantic Ridge shows three essentially similar edifices. These flat-topped volcanoes are aligned north-east to south-west across the lower left-hand side of the image. They are composite volcanoes in so far as they are lumpy features. The lower left-hand volcano has a small central crater developed at its summit. Another cratered flat-topped volcano, to the north-west of the centre of the main cluster, casts a particularly long acoustic shadow and is probably taller than the other volcanoes in the scene.

The contrast variation on the flanks of the volcanoes indicates a roughness on a scale of tens of metres, which is probably a result of the formation of volcanic hummocks or mounds. Typically, the roughness of submarine lava flows is a function of the effusion rate. So for these volcanic edifices, where they appear to be composed of piles of volcanic hummocks, the effusion rates are expected to have been relatively low. At a fine scale, we would expect the volcanoes to comprise piles of pillow lava mounds. Yet as circular features, these edifices have almost certainly evolved from point sources of magma supply and as such represent monogenic features. This conclusion is supported by the occurrence of summit craters on two of the volcanoes.

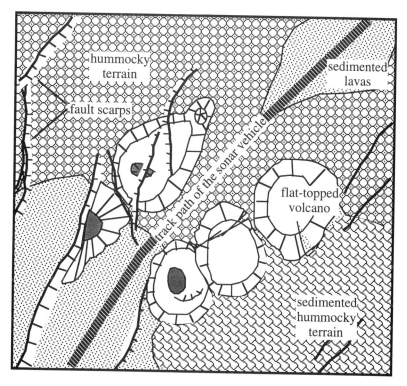

Figure 5.21. Line interpretation of the essential features comprising a volcano cluster on the Mid-Atlantic Ridge, derived from the TOBI sonograph presented in Figure 5.20. The cross-hatched line running diagonally across the centre of the diagram is the track taken by the sonar. The width of the diagram represents a distance of 2.5 km.

Because of their proximity to the sonar vehicle those volcanoes (e.g. to the top left side of the vehicle track) have saturated in backscatter signals. Despite the rough terrain, a good altitude determination of the vehicle has allowed a tight slant-range correction of the image such that the vehicle track appears as a narrow line of disturbance and there is minimal distortion of the image to the port and starboard sides of the vehicle. In this particular image, the side to the lower right-hand of the vehicle track is slightly more blurred than its opposite. This is due to a temporary failure of several elements in the sidescan sonar's starboard transducer array during acquisition. The result of this failure has been an effective decrease in the sonar aperture on the starboard side, causing along-track blurring of the image.

This example demonstrates the importance of knowing the acquisition configuration and conditions during a survey to preclude misinterpretation of the image. Without this knowledge, in this case, it would be an easy mistake to conclude that the seafloor in the lower right-hand side of the scene comprises many short, track-parallel, fissures that have subsequently been draped with thick pelagic sediment. Likewise, the terrain on both sides of the vehicle is also composed of hummocky volcanics. However, the reduced contrast variation for the lava hummocks on the lower left-hand side of the scene is evidence of only a slight sedimented cover.

In contrast to point source volcanic features, another basic building block of oceanic crust is the hummocky volcanic ridge. These features are often half to one kilometre long, a few hundreds of metres wide and up to one hundred metres high. Their importance varies with decreasing spreading rate and probably reflects lower magma fluxes than the flat-topped volcanoes.

5.3.6 Hummocky volcanic ridges

The hummocky ridge imaged by TOBI from the Mid-Atlantic Ridge (Figure 5.22) shows the characteristically heterogeneous nature of the terrain from which it is formed. The scene shows a highly reflective region facing the sonar track, behind which a shadow is cast. From this we can infer that the bright area is a slope facing to the right (i.e. towards the sonar). The elevation of the hummocky ridge, relative to the altitude of the sonar, is sufficient to cast a long shadow. The curvilinear line following the change from reflection to shadow denotes the crest of the feature, which we can determine to be a ridge. The high variation in contrast for the terrain forming the ridge indicates a roughness on a scale of tens of metres. Knowing that this is an image of the axial valley of an active spreading centre, we interpret the rough terrain as multiple eruptive centres that have coalesced to form the hummocky volcanic ridge. Behind the acoustic shadow region, cast by the ridge, is a bright area that has a characteristic speckle and parallel streaks indicative of a steep scarp facing the sonar and covered in scree. Close to the sonar track, a number of bright lobate reflectors are caused by a number of parallel low-elevation curvilinear slopes facing the vehicle. The curved nature of the slopes in this environment means that they are unlikely to be faults. Instead, they are more probably the curved fronts of lobate lava flows that have originated on the hummocky ridge and flowed down and away from its flanks. Despite being in the shadow region on the far side of the ridge, backscatter information is apparent. This is because the dark side slope of the hummocky ridge is close to the grazing angle of the sonar beam and diffraction of the acoustic pulse allows some energy to reach the seafloor. Hence although the backscatter information from the shadow region is attenuated, it is still geologically reliable.

The elongate plan form of the hummocky ridge is a function of fissure-controlled eruption. Often such ridges are formed orthogonal to the spreading direction and indicate the direction of the minimum compressive stress. The predominance of multiple eruptive centres (i.e. haystack edifices), several tens of metres high and in diameter indicates low effusion rates. From this we would expect to find pillow lavas dominating the lava type on hummocky ridges like this one. Because they are formed predominantly from low lava eruption rates, hummocky ridges tend to characterise slow-spreading ridges.

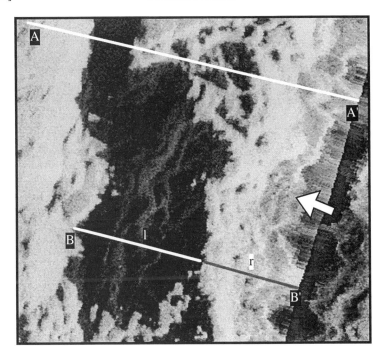

Figure 5.22. Hummocky volcanic ridge, imaged by TOBI 30-kHz sidescan sonar on the Mid-Atlantic Ridge at 3,020 m. The sonar track is shown by the striped region oriented north-east to south-west in the left-hand side of the scene. The arrow shows the ensonification direction. The scene is 2 km wide. The lines A-A' and B-B' indicate the positions of the shape profiles shown in Figure 5.24.

Despite giving information about terrains and features in essentially plan form, sidescan sonar images can also reveal useful information about the vertical profiles of objects. Figure 5.24 shows how the shape across-track can be derived by scaling shape profiles (cf. Figures 5.3 and 5.5) to calculated elevations made from triangulating the shadow length cast by the feature with the altitude and range of that feature to the sonar. From Figure 5.22, the ridge height (h) is derived directly from the shadow length (l), range (r) and vehicle height (H) geometry. This allows the vertical scaling factor for the shape profile, derived by integrating the pixel values along the line B-B', to be calculated. However, because along line A-A the seafloor slopes towards the vehicle behind the hummocky ridge', the shadow cast by the ridge at this point is foreshortened. Therefore to derive the elevation (h') of the ridge along line A-A', h' must be derived proportionally by scaling the elevation at this point to the better constrained elevation along line B-B'. This method is subject to errors where the backscatter law differs significantly between the two profiles, for example where there is a large variation in seafloor texture from one profile to another. In this example, both profiles cross similarly fresh and hummocky volcanic terrain, and so there is no significant change in the backscatter law. These two scaling factors can then be applied to the skyline profile (Figure 5.25).

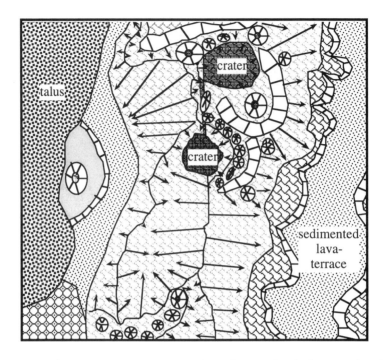

Figure 5.23. Line-drawing interpretation of the volcanic hummocky ridge shown above in Figure 5.15 with the scale. Arrows point downslope, and the more significant volcanic mounds are shown as circles.

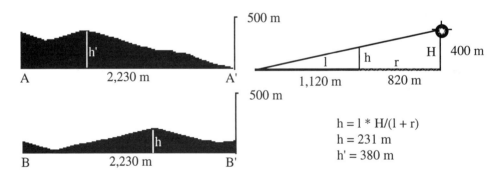

Figure 5.24. Shape profiles derived from integrating the pixel values along lines A-A' and B-B'. The relative vertical scales have been matched by: calculating the heights of the ridge across the two profiles using simple triangulation, knowing the length of the shadows cast at the section lines and the range, and knowing the altitude of the sonar.

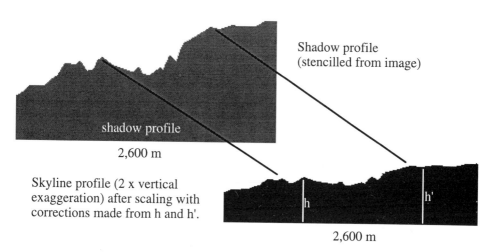

Shadow profile
(stencilled from image)

shadow profile

2,600 m

Skyline profile (2 x vertical
exaggeration) after scaling with
corrections made from h and h'.

2,600 m

Figure 5.25. The skyline profile for the hummocky volcanic ridge has been derived by tracing the outline of the shadow cast by the ridge and then scaling at the two points that correspond to the elevations h and h'. The result is a profile that reveals the shape of the hummocky ridge's silhouette as if it had been viewed from the sonar as it passed.

Hummocky volcanic ridges occur in a number of spreading ridge environments, regardless of the spreading rate. Figure 5.26 illustrates such features imaged by SeaMARC I on the East Pacific Rise. The SeaMARC I sonar is similar to the TOBI system, being deep-towed, relatively high-resolution and with a swath width of 5 km. The image reveals three contrasting textures indicating three principal terrains: the bright mottled texture of the hummocky volcanic ridge, the more subdued mottled texture of sedimented hummocky volcanics and the smooth darker tones of thicker sediment.

Both this example and the one in Figure 5.28 show the merging together of two swaths of processed sidescan data. Known as stencilling, the merging requires the images to be geographically corrected and referenced in a common geographical projection. This can only be achieved when the position of the sonar instrument is known accurately (see Chapter 3 for details).

When interpreting mosaicked images like this, it is important to recognise the ensonified direction for each part of the image. For example, fault scarps facing the sonar and appearing bright as a result, will appear as shadows on the opposite side of the sonar's track. But these scarps will again appear bright where they extend across the image and are ensonified by the next swath of sonar data. Therefore a feature can alternate between bright reflections and shadows as it crosses the mosaic without changing the polarity of its slope.

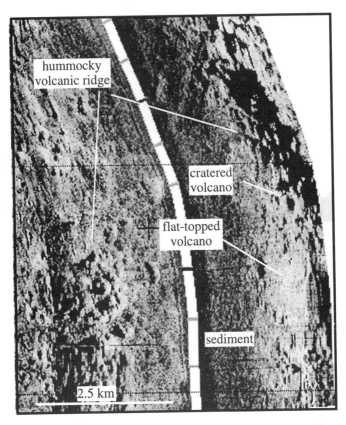

Figure 5.26. SeaMARC I deep-towed sidescan sonar image of hummocky volcanic ridges and terrain on the East Pacific Rise. Note the similarities between this SeaMARC I image of the East Pacific Rise and the TOBI image of the Mid-Atlantic Ridge (Figure 5.22). Both have the same texture, with many small volcanic mounds and a few simple volcanic cones having coalesced to form the hummocky ridges. Courtesy RIDGE WWW Database.

The far left-hand side of the image in Figure 5.26 is the right-hand edge of another parallel swath of SeaMARC I imagery that has been stencilled together with the main area. Care must be taken when interpreting stencilled or mosaicked images to ensure that the ensonification direction for the various regions is realised. For example, in Figure 5.26, the lower left-hand side of the image shows a region of mottled terrain with the part nearest to the sonar track (the white line running top to bottom across the middle of the image) brighter than the rest. In fact, the part of the hummocky terrain furthest from the sonar track belongs to another pass of the sonar and has been ensonified from the left, otherwise it would appear in shadow. Thus there are two identifiable hummocky volcanic ridges (Figure 5.27): one in the lower left of the image

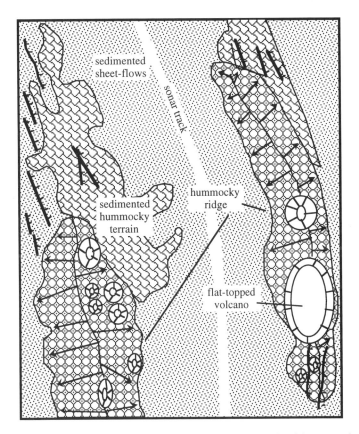

Figure 5.27. Line interpretation of hummocky volcanic ridges on the East Pacific Rise, imaged by the SeaMARC I deep-towed sidescan sonar system. The white line running through the centre of the image is the track taken by the sonar. The scale is the same as for Figure 5.26.

and another forming the right-hand side. The ridge on the right has a small dark circular feature surrounded on the side nearest to the sonar by a bright crescent. This is a cratered conical volcano in the centre of the hummocky ridge. A larger circular feature, that is also bright, lies just to the south of the cratered volcano. This feature also has a brighter crescent forming its rim closest to the sonar and a shadow on its far side. We interpret this feature to be a larger flat-topped volcano. The dark and light linear features are fault scarps facing away and towards the sonar respectively.

5.3.7 Axial Volcanic Ridges

Axial volcanic ridges are common on slow- to medium-spreading ridges. Formed as elongate neovolcanic features, 10 to 50 km long by 3 to 10 km wide and 50 to 200 m high, they are an essential unit of volcanic construction of the ocean crust. They are formed by coalescence of individual volcanic units such as conical and flat-topped volcanoes, hummocky volcanic ridges and a variety of lava flow types (Parson et al., 1993).

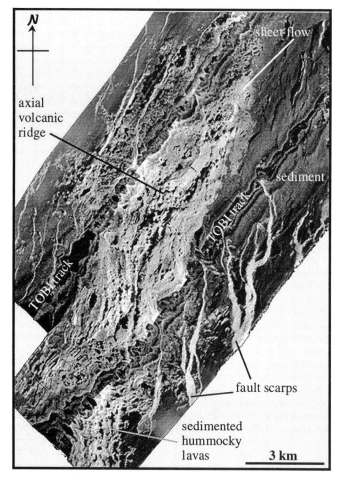

Figure 5.28. Stencilled and mosaicked TOBI imagery of an axial volcanic ridge on the Reykjanes Ridge, 1,000 km south-west of Iceland (after Murton and Parson, 1993). The bright almond-shaped area is the volcanic feature and the darker surrounding material is a thick sediment cover. The curvilinear bright and dark features are fault scarps.

This particular image (Figure 5.28) is an excellent example of contrasting seafloor acoustic facies: the bright regions being rough volcanic terrain, and the dark regions being areas covered with a thick pelagic sediment drape. The smooth pelagic sediment has two effects on the sonar beam. Its smooth surface reflects the beam away from the sonar, and its softness absorbs and attenuates the beam. Thus the resulting acoustic echoes from pelagic sediment are weak.

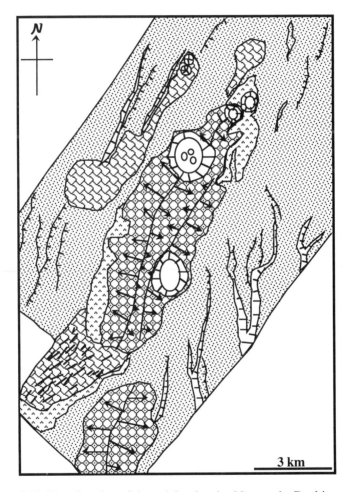

Figure 5.29. Line drawing of the axial volcanic ridge on the Reykjanes Ridge south-west of Iceland. There are three quite distinct features: the fresh to slightly covered volcanic terrain forming the AVR and containing several flat-topped and conical volcanoes, the monotonous sedimented areas and the curvilinear faults.

In this example, the almond-shaped region of bright mottled texture is the axial volcanic ridge (AVR) which comprises hummocky volcanic terrain with a few larger volcanic edifices. Smooth bright reflective areas with dentritic margins at the northern tip and western side of the AVR are sheet flows, confirmed by rocks recovered from these areas. The darker smooth areas surrounding the brighter reflecting volcanic terrain of the AVR are pelagic sediment-draped regions.

The hummocky areas, whose reflectivity is intermediate between the AVR and pelagic sediment drape, are interpreted as partially sedimented hummocky volcanics (Figure 5.29). To the south of the AVR, such sedimented volcanics are dissected by a dense number of short lineaments. These appear as dark lines and are probably either fissures or small faults facing away from the sonar vehicle. Bright curvilinear reflectors on the margins of the scene are fault scarps facing the sonar. Dark, curvilinear features in the north-east of the scene are faults facing away from the sonar. The AVR onlaps the sedimented and faulted terrain and is thus younger.

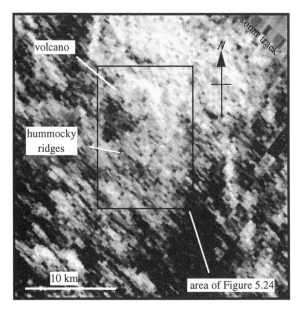

Figure 5.30. GLORIA 6.5-kHz sidescan sonar image of the Mid-Atlantic Ridge south-west of the Azores islands (copyright Southampton Oceanography Centre, UK). The coarse-resolution imagery has a pixel size of about 100 m^2 in the mid-range and a swath width of 50 km in deep water. Note the along-track blurring of targets caused by the wide beam angle. GLORIA is a regional mapping sidescan sonar and its successor, GLORIA-B, has paired transducers producing additional phase-difference swath bathymetry.

Recent studies of axial volcanic ridges (AVRs) have shown that they appear to have a variety of ages, despite being within the axial valley of active spreading ridges. This variation has been ascribed to the AVRs having a cyclic history of volcanism and tectonism causing them to grow and then degrade (Murton and Parson 1993). The Reykjanes Ridge, where AVRs were first fully recognised (Searle and Laughton, 1981), is a slow-spreading ridge. However, being close to the Icelandic hot-spot, it has a high volcanic productivity. Nevertheless, these volcanic features have been described from other ridge systems elsewhere, regardless of the spreading rate or volcanic supply, and are believed to play an important role in the construction of the oceanic crust. Geochemical studies of these features have shown that they have internally consistent chemistry indicating evolution from a common magma supply (i.e. co-genetic) but that adjacent ridges differ and are hence isolated individual magmatic systems (Taylor et al., 1995)

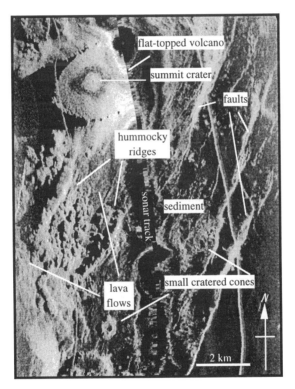

Figure 5.31. TOBI sidescan sonar image of the Mid-Atlantic Ridge southwest of the Azores islands (courtesy of L. M. Parson, Southampton Oceanography Centre, UK). This is the same area as indicated by the box drawn on the GLORIA image shown in Figure 5.30. The huge differences in detail resolved by the two sonar systems mean that the data are collected and used for different purposes. TOBI imagery allows more local studies, for example of the axis of a spreading ridge, whereas GLORIA imagery enables large-scale regional views.

As the scale of observations increases to encompass different volcanic features on sections of spreading ridges, so the type of sonar typically employed changes. Instead of making high-resolution surveys with deep-towed sonars, surface-towed low-frequency systems are employed that are towed at high speed and with large swath widths. However, the advantage of speed of survey is offset by the disadvantage of poorer resolution. Comparison between Figures 5.30 and 5.31 shows the difference in detail revealed by both GLORIA and TOBI images of the same feature on the Mid-Atlantic Ridge.

The abrupt change in resolution from the 6.5-kHz GLORIA imagery to deep-towed sidescan sonars like the 30-kHz TOBI and SeaMARC I makes it very difficult for geologists to recognise common features. A volcano, which is just identifiable on the GLORIA image is revealed in great detail by TOBI. Similarly, faults and terrain variation seen on the TOBI image are just not resolvable from the GLORIA image. The advantage of GLORIA imagery is discussed later, where many parallel passes of the sonar, made at a survey ship's cruising speed, are mosaicked together to generate regional images of large areas of the seafloor.

5.4 TECTONIC FEATURES

5.4.1 General Tectonic Trends at Mid-Ocean Ridges

Deformation at oceanic spreading centres is predominately by brittle extension and shear. Extensional fault patterns on spreading ridges generally follow two trends: those that are orthogonal to the plate extension direction, and those parallel to the plate boundary. The first of these is concentrated within the floor of the axis: an axial valley if it is a slow- to medium-spreading ridge, and a crest if it is a fast-spreading ridge. These tectonic structures are largely extensional in character and form either fissures (with no vertical offset across them) or fault scarps (with varying amounts of vertical offset). While the majority of fault scarps face in towards the centre of the ridge axis, some faults face outwards. The structures that are parallel to the trend of the plate boundary will have the same orientation as those on the floor of the axis only where the ridge axis is orthogonal to the spreading direction. For most ridges, this is a close approximation, but with slower spreading rates, the ridge axes tend to have a greater number of oblique segments.

Depending on the orientation of the faults with respect to the sonar, they either appear as bright linear reflectors where they face the sonar, or cast linear shadows where they face away from the instrument. Often faults are seen to link together in complex ways, or, especially for larger faults, terminate in complex splays. They are also effected by interaction with topography, such as volcanoes and axial volcanic ridges. Large fault scarps, which characterise the axial valley walls on slow- to medium-spreading ridges, are often covered in mass-wasting sediments such as scree and sometimes serpentinites. These axial valley wall faults act to uplift the seafloor as it moves away from the ridge axis, rather like an escalator. Earthquake activity on these faults often displaces large

Colour Section

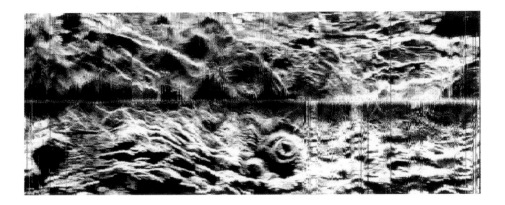

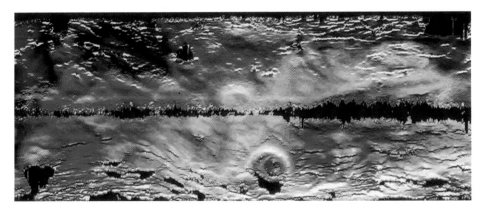

Plate 1. Three panels showing imagery and bathymetry data from the DSL-120 sidescan sonar. Upper panel is a swath of backscatter, 1 km wide. The prominent circular feature in the lower middle of the swath is the TAG hydrothermal mound. Linear bright and dark features are faults and fissures which cut across the seafloor. The middle panel shows the bathymetry at a ~2 m^2 resolution (colour coded from red-shallow to mauve-deep). The lower panel represents the grey-shaded relief, artificially illuminated from the top to the bottom of the image. (Courtesy S. Humphris and M. Kleinrock, Woods Hole Oceanographic Institute, USA).

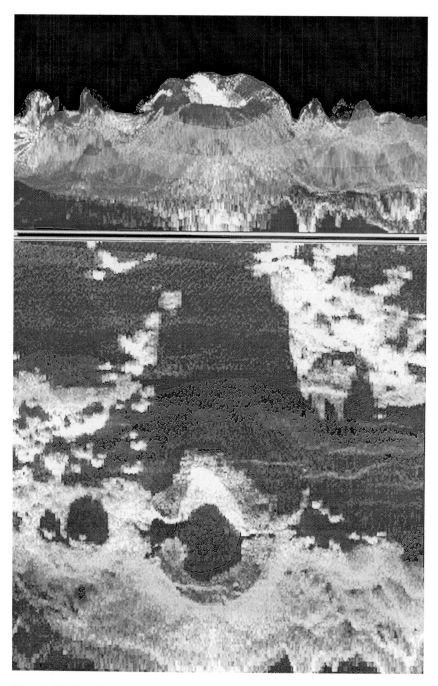

Plate 2. TOBI sidescan sonar image of the W-Seamount, 29°N on the Mid-Atlantic Ridge. The upper panel shows a prespective projection of sidescan imagery draped over multibeam bathymetry, collected during a previous survey and co-registered with the sonar data. The lower panel shows the same merged TOBI imagery and multibeam bathymetry, colour-coded (red = shallow to mauve = deep), but presented in plan view.

Plate 3. Plan view of the Indian Ridge Triple Junction: multibeam bathymetry has been co-registered with GLORIA sidescan sonar imagery (depths are colour coded: red-shallow to mauve-deep). This is an example of the fusion between two regional-scale sonar data sets. (Courtesy T. L. LeBas, Southampton Oceanography Centre, UK.)

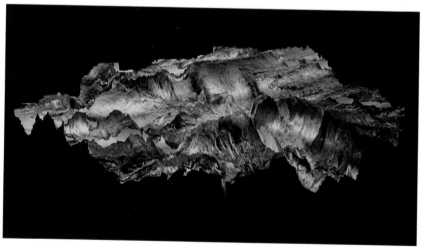

Plate 4. Perspective view of merged GLORIA sidescan sonar imagery and multibeam bathymetry from the Indian Ridge Triple Junction. This style of displaying different data sets allows an interpretation of the backscatter in the context of the local slopes.

Plate 5a. Perspective projection (from the South) of co-registered and merged TOBI sidescan sonar imagery and multibeam bathymetry at 29°N on the Mid-Atlantic Ridge. Unlike GLORIA, the co-registration of the higher-resolution TOBI imagery requires sophisticated navigation processing for the deep-towed vehicle. The colour-coding for depths are: red shallow to mauve deep.

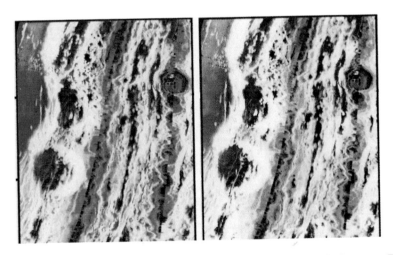

Plate 5b. Plan view of merged TOBI sidescan sonar and multibeam bathymetry. The left-hand panel shows the merged datasets (colour coded for depth: red shallow to mauve deep). The right-hand panel shows the backscatter intensity from TOBI, but without any bathymetry information.

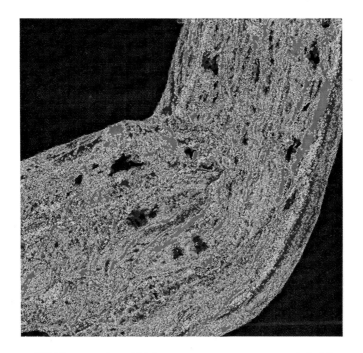

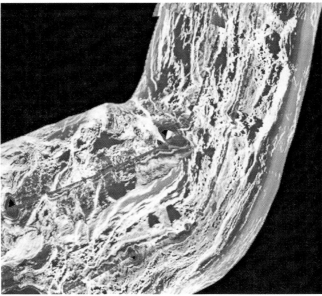

Plate 6. The upper panel shows mosaicked TOBI swaths of a non-transform discontinuity from the Mid-Atlantic Ridge. These data are classified on the basis of first-order statistics (mean, maximum, minimum, and standard deviation computed on 3x3 neighbourhood). The classification, shown in the lower panel, is based on ground-truthed areas from which the critical statistical envelopes describing each class of terrain have been extracted. The method displays five terrain types: talus slopes (red), pelagic sediment (blue), chaotic sediment (green), volcanics (yellow) and shadows and areas without data (black).

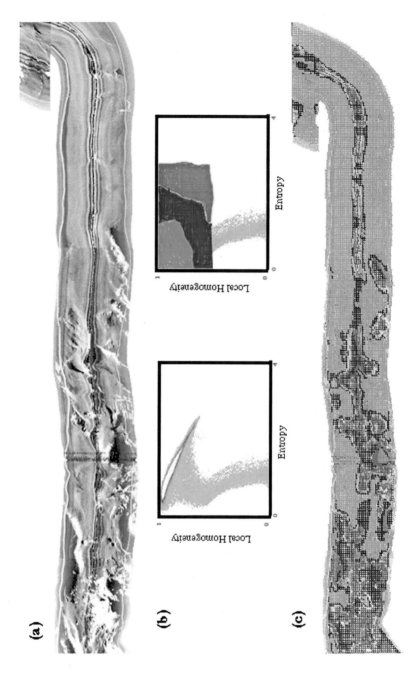

Plate 7. Panel (a) shows the original TOBI image of a mid-ocean ridge discontinuity. Swath width is 6 km, higher backscatter values are brighter. Panel (b) shows the distribution of the main geologic provinces in the {entropy; homogeneity} provinces. Textures from neo-volcanic provinces are represented in red, tectonised areas in blue, sedimented areas in green. Panel (c) shows the map of acoustic textures superposed on top of the imagery. Textural analysis allows the automatic mapping of geological provinces by extrapolating from a few ground-truthed areas. (From Blondel, 1996)

volumes of pelagic sediment that form mud slides that cover the slopes of the fault scarps and areas of the surrounding axial floor.

Shearing of the oceanic crust is most prevalent in transform faults. These are strike-slip systems where lateral offsets in the spreading ridge axes cause adjacent sections of the newly formed oceanic crust to slide past one another. These regions are host to quite complex tectonics, and a general lack of volcanism within them means that they accumulate significant thicknesses of pelagic sediment. These sediments often become locally distorted and contorted by the fault activity in to complicated structures.

5.4.2 Single Faults and Fissures

Where faults face the sonar, their steep slopes return strong echoes. The high intensity of these echoes combined with the linear shape of the features allows us to identify the targets as being probable fault scarps. To illustrate this process, and to show the different appearance of the same fault sets imaged by two different sonars, a series of panels is displayed in Figures 5.32 and 5.33.

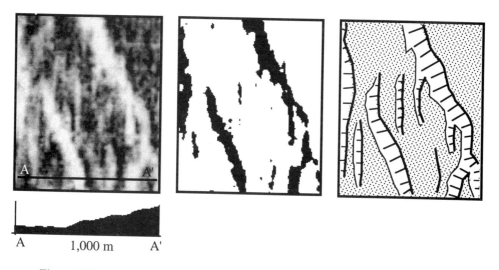

Figure 5.32. SeaMARC II image of a series of faults facing the sonar (i.e. to the left) from the Reykjanes Ridge (courtesy S. Shor, University of Hawaii, USA). The original image is shown in the left-hand panel, along with a shape profile showing the stepped nature of the faults. The middle panel shows the strongest 15% of echoes corresponding to the steepest slopes facing the sonar. The right-hand panel shows the bright linear features interpreted as faults.

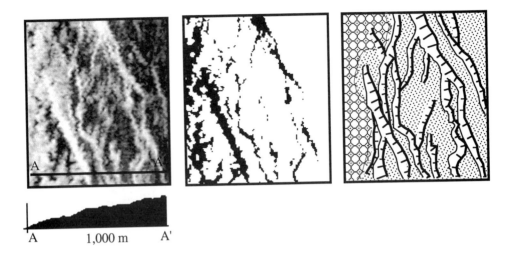

Figure 5.33. The first panel shows a TOBI image of the same scene as the one shown in Figure 5.32, along with a shape profile that reveals more details of the faulting. The middle panel is a threshold slice of the strongest 15% of echoes, and the right-hand panel shows an interpretation of the fault pattern.

In each figure, the first panel shows the original sonar image with a shape profile drawn across the faults to give an impression of the step-like nature of these features. The step-like shape of axial floor faults is exemplified in a TOBI image from the Mid-Atlantic Ridge (Figure 5.4). A shape profile, generated from this imagery, is shown in Figure 5.5. The second panel shows the result of a process known as thresholding, in which only the strongest 15% of echoes are displayed (as black pixels). This identifies the steepest slopes facing towards the sonar, yet does not differentiate slopes that are volcanic from those that are tectonic. To do this, only those steep slopes that form part of a linear feature are interpreted as faults. The third panel shows the final result of the process in which the tectonic structures are determined. It relies on the interpreter recognising a pattern of bright linear features that disrupt the scene, either by cutting across terrains or of setting other features such as volcanoes or hummocky volcanic ridges. Noisy, discontinuous, or bitty reflectors, especially those forming the left-hand side of the middle panel in Figure 5.26, are interpreted as the flanks of coalesced volcanic hummocks and hence not regarded as being of tectonic origin.

The shape profiles from the faulted region are particularly interesting because they show that the area behind each fault is tilted away from the fault scarp. This phenomenon, known as back-tilting, is common for faults in an extensional environment. The pattern of faulting is also interesting because it shows merging of large faults, smaller fault splays and linkage between larger and smaller faults.

5.4.3 Fault Scarps

In dictionaries of geology, faults are defined as "fracture surface or zones along which appreciable displacement has taken place". The definition of the displacement is volontarily vague, as it can happen on scales ranging from mesoscopic (tens to hundreds of metres) to megascopic (several kilometres or tens of kilometres). Fault scarps present themselves on sonar imagery in the same way fissures do. Scarps facing toward the sonar will be bright, the more so as the local angle of incidence will be high (i.e. the more perpendicular to the acoustic wave they are). And fault scarps facing away from the sonar will be darker or will not reflect any energy.

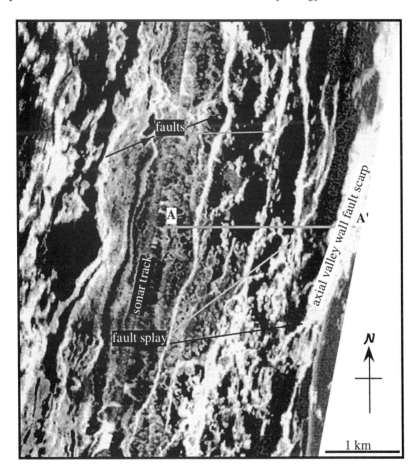

Figure 5.34. Section of the eastern half of the axial valley floor of the Mid-Atlantic Ridge. Bright linear reflectors are fault scarps ranging from a few metres to tens of metres high and facing to the West. The shape profile A-A' reveals both the different heights, slopes and spacings of the faults (see Figure 5.35). It also shows the back-tilting of fault blocks characteristic of mid-ocean ridge terrain.

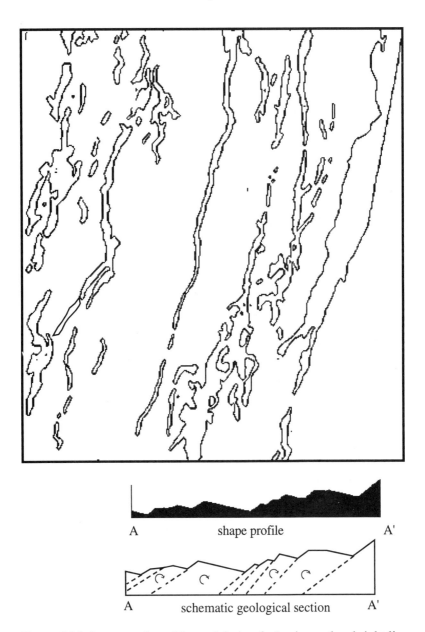

Figure 5.35. Interpretation of inward facing faults, imaged as bright linear reflectors on Figure 5.34. The interpretation is made from by contouring a binary version of a threshold slice of the highest 15% of dn values and for which only the linear features are preserved. Faults facing away from the sonar, and hence appearing as shadows, are not shown since those scarps can only be inferred.

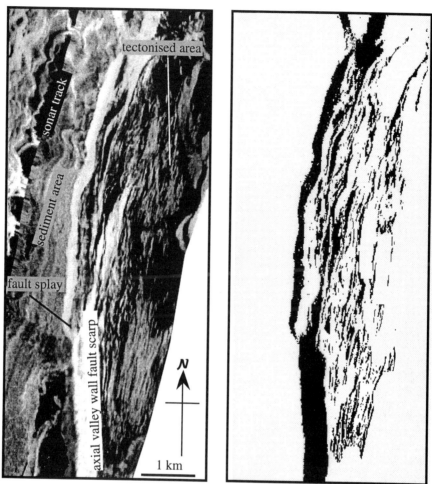

Figure 5.36. The left-hand panel shows a TOBI image of major axial valley wall fault splays, bifurcations and tectonised terrain. A semi-automatic interpretation of the structural components of the scene is shown in the right-hand panel.

Because they are much larger than fissures, fault scarps are not always purely tectonic structures. Some of them have been reaffected by tectonic activity, and support a host of smaller faults and fissures. Others have been overlain by volcanic flows or sediments. The examples of this section show these different types in detail, and in different geological settings.

In the TOBI image of a major, west-facing, axial valley wall fault scarp on the Mid-Atlantic Ridge (Figure 5.36), the fault scarps face west. Behind the major scarps, a series of sub-parallel thin bright and dark lines closely spaced together are regions of fissured terrain and minor faults that face both inward and outward from the centre of the ridge axis. The fault scarp has a series of bifurcations, the most significant being one-third of the way from the bottom of the image. Beyond the bifurcations, the faults splay into a number of minor faults fissures and fractures that dominate the right-hand side of the image. The left-hand side comprises the nadir region of the sonar vehicle. Smoother grey tones indicate pelagic sediment cover. The top left-hand corner of the image contains bright reflectors with a lobate shape and are probably lava flow fronts.

The main structural components are shown in the right-hand panel of Figure 5.36. This image has been made by a combination of threshold slices of the lowest 5% and highest 15% of pixel values, displayed as black, and filtered to remove the less linear features. The region on the lowermost left-hand corner of the image was selectively enhanced before thresholding to take account of the reduced contrast in this region. The cause of the lower contrast is probably a localised patch of sediment that was redeposited from further up the axial valley wall. The resulting image is a reasonable representation of the structure in the scene with contribution from the bright faults facing the sonar, and the dark fissures, being defined only by their shadow regardless of the ensonification direction. It is quite possible to combine the threshold slices in such a way that the dark linear features appear in one colour while the bright linear features appear in another. This can differentiate between those inward-facing faults and those that face outward as well as the fissures.

The deep-tow sidescan sonar SeaMARC I image of the East Pacific Rise (Figure 5.37) shows the contrast between the relatively low relief of this fast-spreading ridge compared with the high relief of slower-spreading ridges such as the Mid-Atlantic Ridge. The image reveals a relatively smooth terrain that has areas of mottled texture and linear bright and dark features. The vehicle track is indicated by the white line passing through the centre of the image with black ticks following 30 minutes of data acquisition time. Given that this is an image of a fast-spreading ridge where magma supply is large, the smooth areas are almost certainly sheet flows. The mottled areas are mounds and piles of pillow lavas which in the centre right-hand side of the image have coalesced to form a low-relief volcano. The centre of the volcano contains a circular shadow that is probably a central crater. Linear features running across the scene are faults, bright ones being those facing the sonar and dark ones being both fissures and faults facing away from the sonar. A characteristic feature of fast-spreading ridges like the East Pacific Rise is the prevalence of fault scarps that face both towards and away from the ridge axis. Similarly, fast-spreading ridges have lower fault throws (i.e. lower scarps) than for slow-spreading ridges, the faults tend to bifurcate less and axial valleys and valley walls tend to be absent.

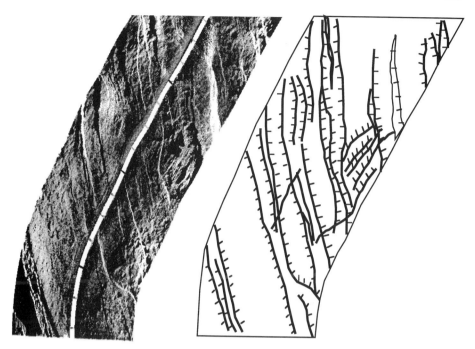

Figure 5.37. The left-hand panel shows a 30-kHz SeaMARC I image of the axial valley of the East Pacific Rise. Fault splays and bifurcations are shown in the right-hand panel where they have been interpreted by hand. Ticks denote the downward slope of the fault scarps. This sonar has a swath width of 5 km and the nadir region is shown by the thick white line. Courtesy RIDGE WWW Database.

5.4.4 Tectonic Features at Transform and Non-Transform Offsets

Many slow- and medium-spreading ridges are characterised by significant along-axis segmentation, either in the form of transform faults which offset the axis by tens of kilometres, or by non-transform discontinuities which are often less than 10 km long. Based on analysis of bathymetry and magnetics, spreading centre discontinuities have been separated into different classes. The longest lived (> 2 Ma) and the largest (20-30 km) of these offsets correspond to transform faults, or first-order discontinuities. Along the Mid-Atlantic Ridge, they offset spreading segments at intervals of 200-800 km. They often correspond to narrow strike-slip fault zones.

Non-transform offsets, or discontinuities, are another typical feature along the Mid Atlantic Ridge, in particular between the Kane and Atlantis Fracture Zones. They are defined by displacements of the along-strike bathymetric highs and of the highly magnetic expression of the ridge axis. They can be traced off-axis as a series of interlinked basins and breaks in the otherwise linear ridge parallel magnetic anomalies. Non-transform discontinuities are also characterised by positive gravity highs, both on- and off-axis. They are important because of the insights they give about the evolution of spreading segments. With its high resolution, sidescan sonar imagery gives access to smaller-scale processes than could be observed by conventional geophysics.

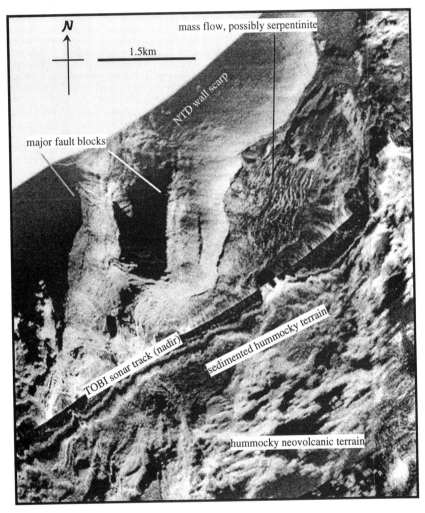

Figure 5.38. A TOBI image of the western flanking wall of a non-transform discontinuity on the Mid-Atlantic Ridge. The steep wall (top left) returns strong echoes and the streaks running east-west across the scarps are scree that has eroded from the outcropping rock. To the right of the scarps a streaky dark mass that widens from the top to the bottom of the scene is a serpentinite mass flow that has travelled downslope. The scene to the right of the sonar track is dominated by sedimented hummocky volcanic terrain.

In Figure 5.38, the TOBI image covers mainly the western wall of a non-transform discontinuity on the Mid-Atlantic Ridge. The elevation of the nearly shear wall is between 500 m and 1,200 m and represents some of the most dramatic topography in the oceans. A line interpretation of the image is given in Figure 5.39 to accompany the explanation of the scene and its geology.

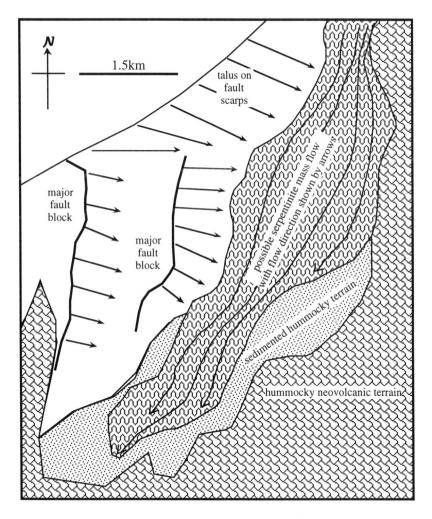

Figure 5.39. Line drawing interpretation of the non-transform discontinuity shown in Figure 5.38. The major fault scarps are shown as white with the arrows depicting the downslope direction illustrated by the streaky scree deposits. The mass-flow serpentinite deposit is a wavy annotation and the arrows indicate the direction of flow derived from the flow's streaky texture.

The bright echoes are strong returns from fault scarps facing the sonar. The shadows behind the scarps indicates that they are upstanding and are probably faulted blocks. The trend of the faults. derived from the junction between the bright returns and the shadows, is north-south. The streaky texture to the scarps is caused by scree deposited on the slopes and derived by downslope erosion of the rock exposed by the faulting. This exposed rock probably forms the brightest echoes which occur as a band running along the junction between the slope and shadow of the middle scarp. To the right of the fault scarps, a dark, streaky mass of material is seen. This material widens from an apex in the top right-hand corner of the scene towards the middle left. The lighter and darker streaks are probably caused by flow in the material, and are especially apparent to the south of a number of small hummocks that cuts across the flow. From sampling, we know that the flow comprises serpentinite which is a hydrated form of peridotite that usually occurs at the base of the oceanic crust (Cann et al., 1992). In this environment, where there is large-scale vertical uplift in the non-transform discontinuity, the ultramafics have probably been brought to the surface by faulting where sea water has hydrated it to become serpentinite. As a material, serpentinite is also known to move downslope relatively easily as a semi-congealed mass flow. The southern side of the sonar image is dominated by smooth, grey, mottled textures that indicate sediment-covered volcanic terrain.

5.5 HYDROTHERMAL FEATURES

The large energy fluxes that occur at volcanically active mid-ocean ridges release huge amounts of heat. This drives water circulation from the overlying water column deep into the oceanic crust, where it is heated up to boiling point. On its return to the surface of the crust, the hot water leaches metals out of the wall rocks and becomes a super-saturated chemical cocktail. Where this fluid exits the crust, hydrothermal vents are formed. Called "black smokers", they expel mineral-rich waters approaching 360°C (Figure 5.40) and release tonnes of these metals into the ocean each year. .

Figure 5.40. Photographs of hydrothermal "black smokers" on the Broken Spur Vent Field at 29°N on the Mid-Atlantic Ridge (photographed from the DSV *ALVIN* by B. J. Murton, 1993).

5.5.1 Hydrothermal Mounds

The smoke seen on visual images of hydrothermal vents corresponds to the precipitation of sulphide and metal oxide.They set down close to the vents, forming valuable mineral deposits. These deposits of mixed sulphide and anhydrite build chimneys, mounds and metal-rich sediment ponds several tens of metres high and hundreds of metres in diameter.

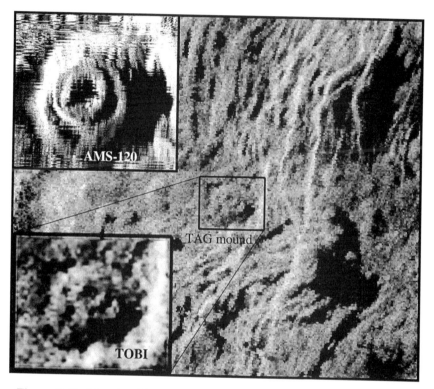

Figure 5.41. Mosaic of imagery of the TAG hydrothermal mound, at 26°N on the Mid-Atlantic Ridge. The main scene is a TOBI sonograph (geographically registered and ensonified from the left) of the western side of the axial valley and is approximately 2.5 km square (courtesy of J. Cann, University of Leeds, UK). The TAG mound is the button-shaped feature in the centre. Inset bottom left contains a times two enlargement and contrast equalised detail of the mound itself. Note the concentric circular nature of the mound with a small conical high to the left of middle. Inset top left is the same area as the TOBI detail of the TAG mound, ensonified from the left by the AMS-120 sonar (courtesy of S. Humphris and M. Kleinrock., Woods Hole Oceanographic Institute, USA). The image is not geographically registered and is rotated by about 45° counterclockwise relative to the TOBI view.

Because of their relatively small size, hydrothermal vents are mostly studied visually. However, high-resolution deep-towed sidescan sonars like the DSL-120 are able to image individual features. The only confirmed image of an active hydrothermal mound on the Mid-Atlantic Ridge is of the TAG mound (Figure 5.41). This sulphide and anhydrite construct is the largest known single hydrothermal deposit in the ocean, and has been the subject of considerable research. It has been imaged by two sidescan sonar systems: TOBI (Cann, unpublished data) and the DSL-120 (Kleinrock and Humphris, 1996). DSL-120 imagery, combined with high-resolution bathymetry, is shown in Plate 1 (in the colour section). The DSL-120 system is a deep-tow sonar that acquires both imagery and phase difference bathymetry at 120-kHz. Its higher frequency and acquisition parameters allow a greater resolution than the TOBI system. Comparison between the TOBI and DSL-120 images reveals the difference in spatial resolution of the two systems. The striped appearance of the DSL-120 image is a result of poor altitude determination of the sonar.

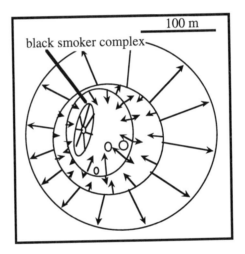

Figure 5.42. Line-drawing interpretation of the TAG hydrothermal mound, at 26°N on the Mid-Atlantic Ridge. The concentric rings are the crests of circular ridges with arrows pointing downslope. The centre of the mound has a large edifice and some smaller lumps which are probably hydrothermal chimneys.

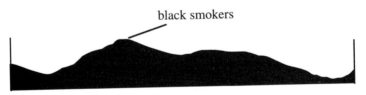

Figure 5.43. Unscaled shape profile from the DSL-120 image of the TAG mound, which can be compared with the profile drawn by observers on board the deep submergence vehicle ALVIN (Figure 5.44).

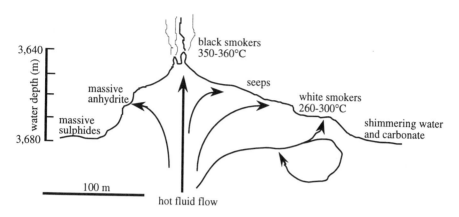

Figure 5.44. Schematic cross-section of the active TAG mound compiled from many manned submersible dives on the site (Rona, 1986). Arrows show the fluid transport direction inferred from the mineralogy and chemistry of the deposits and hydrothermal fluids.

The hydrothermal deposit that forms the TAG mound is a precipitate of iron and copper sulphides mixed with anhydrite. The metalliferous content of the mound is the result of fall-out from the hot fluid discharged from black smokers at the top of the mound. These superheated fluids escape from the mound at temperatures of around 360°C and contain dissolved metals, methane, helium and sulphur. On mixing with the ~2°C ocean-bottom water above the mound, the fluids precipitate their metal content which then forms a plume of microscopic particles. The rain from this plume, along with collapsing sulphide chimneys and precipitation internally of sulphide and anhydrite, builds up the mound structure. Periodic waxing and waning of the hydrothermal activity allows the anhydrite to dissolve and reform, giving the mound its characteristic shape of concentric-rings (Figures 5.42 and 5.43). Internally, seawater and hot hydrothermal fluid circulate and mix, escaping as fluids with different temperatures and compositions (Figure 5.44)

5.5.2 Hydrothermal Structures

Individual hydrothermal constructs are typically small mounds and chimney structures. Although the Mid-Atlantic Ridge is dominated by larger mounds, the faster-spreading East Pacific Rise and Juan de Fuca Ridge are characterised by chimneys. Because hydrothermal chimneys are generally smaller features than the mounds, they can only be imaged acoustically by higher-frequency systems. On the Juan de Fuca Ridge, the DSL-120 sidescan sonar has imaged a number of features that submersible dives have found to be hydrothermal chimneys and mounds (Figure 5.45). Such chimneys are common on this fast-spreading ridge where they are concentrated along, or adjacent to, a central axial graben. The graben, known as an axial caldera, is formed by collapse of a linear high-level magma chamber. The fresh lavas erupted from this feature provide much of the heat that supplies the hydrothermal systems. High magma flux at these ridges causes the hydrothermal deposits to become rapidly buried. Thus the development of large mounds such as the TAG mound is inhibited.

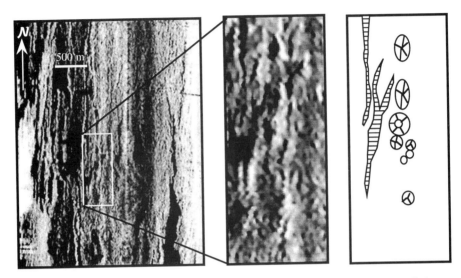

Figure 5.45. Left-hand panel: DSL-120 image of the Hulk hydrothermal site on the Juan de Fuca Ridge, north-east Pacific. The linear features are fissures and faults, with the long linear axial summit graben lying just to the left of the inset square. The middle panel shows an enlargement of the Hulk site and the right-hand panel is a line interpretation of the major fissures (hatched linear features) and mounds (circles).

5.6 REGIONAL IMAGERY

5.6.1 West Lau Spreading Centre

GLORIA imagery, like the one shown in Figure 5.46, is particularly apt at revealing regional-scale geological environments. The entire US EEZ (Exclusive Economic Zone) was mapped by GLORIA in the 1980s, and included the seafloor surrounding the off-shore state of Hawaii. Figure 5.46 shows the complex configuration of oceanic spreading centres that have formed the crust of the Lau Basin, a back-arc basin in the south-western Pacific Ocean. The image is a mosaic of several GLORIA passes.

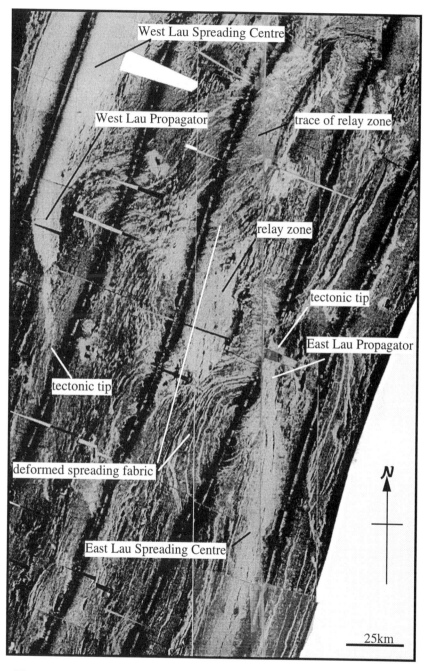

Figure 5.46. Regional mosaic of GLORIA imagery for the northern Lau Basin (Parson et al., 1992).

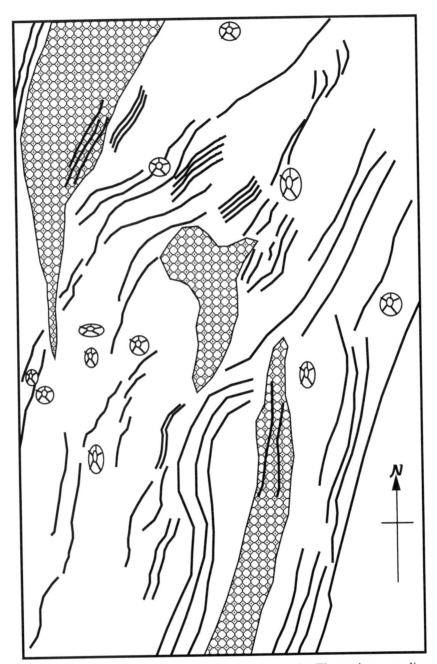

Figure 5.47. Interpretation of the GLORIA mosaic. The major spreading fabric is shown as thick lines, the spreading ridges are shown with a circular patterned fill and individual conical volcanoes are shown by circles.

In the interpretation of the GLORIA mosaic (Figure 5.47), the bright reflecting areas are rough terrain that is characteristic of the neovolcanic nature of oceanic spreading ridges. The darker backscattering areas are those regions covered in smooth sediment. Linear features, both bright and dark, are the major tectonic fabric that results from the action of oceanic spreading. Circular features are individual submarine volcanoes that are both part of the spreading ridge and the volcanic arc that has formed as a result of the melting of the subducting Pacific Plate.

The configuration of the spreading ridges in the Lau basin is particularly complex, resulting from the highly dynamic tectonic environment established behind the Tonga-Kermadec Trench, the world's fastest subducting collision zone. The entire Lau Basin has opened up in the past 5 Ma in the shape of huge "Y". The northern part of the basin is opening faster than the southern part. As a result, the spreading ridges are propagating southward, unzipping the crust as they pass. This mechanism causes two or more ridges to be present, the more northerly growing southward at the expense of the more southerly one. In between, a smaller ephemeral ridge, called a relay zone, moves south with the overlapping ridges. Regional stresses between the overlapping ridge tips causes them to bend towards one another. The passing of the propagator distorts the otherwise ridge-parallel spreading fabric, causing curved faults to appear.

5.6.2 Central Indian Ocean Triple Junction

A major feature of plate tectonics is the triple junction; the loci of three types of plate boundary. The Indian Ocean is perhaps one of the more difficult regions to study because of its remoteness. Yet it is host to a spectacular ridge-ridge-ridge triple junction where three mid-ocean ridges with greatly different spreading rates meet. The triple junction has a history of migration, leaving a complex trail of crust formed at the different ridges. GLORIA sidescan sonar imagery alone (Figure 5.48) reveals much information about the spreading fabric of the crust, the location of the major fault scarps and areas of fresher lavas. However, the region of the triple junction is extremely complex and other data sets are required to interpret the geology (Figure 5.49). These include: bathymetry, magnetics and gravity data. Figure 5.50 shows the GLORIA imagery draped over the bathymetry (from a multibeam echo-sounder). Such overlain data sets greatly enhance the geologist's ability to interpret the scene.

Combining data sets is often better done by creating colour views. The GLORIA image shown in Figure 5.48 is colour-coded for depth, and the brightness is a function of the sidescan backscatter intensity (Plate 3 in the colour section). Similarly, the draped perspective view is enhanced by colour coding the brightness for depth, and this is shown on Plate 4 in the colour section. It is not only depth that can be used to colour-code sidescan sonar images; other data sets such as magnetics and gravity can also be used. This method of data combination is explored further in Chapter 10 where Geographic Information Systems are introduced.

Figure 5.48. GLORIA imagery of the Central Indian Ocean Triple Junction (courtesy of L. M. Parson and T. LeBas, Southampton Oceanography Centre, UK). Composed of five GLORIA passes oriented north to south, the imagery is dominated by the spreading fabrics of three ridges. A regular linear fabric from the fast-spreading South-East Indian Ocean Ridge forms the south-eastern half of the scene. The northern region is dominated by a more irregular fabric from the slower-spreading South-East Indian Ocean Ridge. A triangular-shaped wedge of bright crust are large elevation scarps formed by the ultra-slow spreading South-West Indian Ridge.

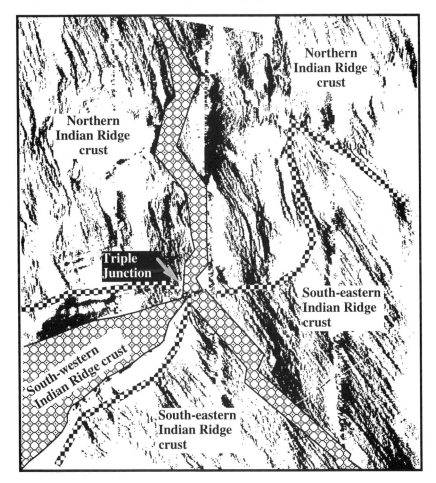

Figure 5.49. Line interpretation of the GLORIA imagery of the Central Indian Ocean Triple Junction. The line interpretation has been formed by thresholding the image to show the dominant linear spreading fabrics. The loci of the spreading ridges (circular pattern) is chosen on the basis of the brightest reflections (i.e. freshest lavas), the mid-point between inward- and outward- facing faults, the deepest bathymetry, and the youngest magnetic lineaments. The boundaries between the crust formed by the ridges (chequered lines), is chosen on the basis of changes in the style of spreading fabrics and magnetic lineations.

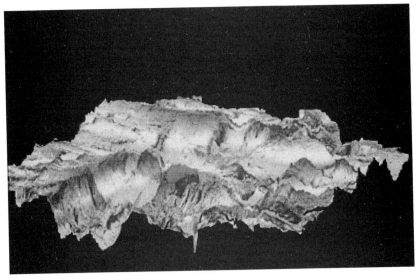

Figure 5.50. GLORIA imagery draped over multibeam bathymetry and projected as a perspective view from the south (courtesy T. LeBas, Southampton Oceanography Centre, UK). The trough to the left is the ultra-slow spreading South-West Indian Ocean Ridge where magmatic flux is so low that there is almost no recent volcanism. The trough to the right is the South-East Indian ridge, an ultra-fast spreading rate ridge similar in style to the East Pacific Rise. The trough trending to the left is the medium-spreading Central or North Indian Ridge, and is similar to the Mid-Atlantic Ridge.

5.7 FURTHER READING

- ### About mid-ocean ridges:

ARCYANA, "FAMOUS: Photographic atlas of the Mid-Atlantic Ridge: Rift and transform fault at 3000 meters depth", Paris: Gauthiers-Villars & Cnexo, 126 pp., 1978

Cullen, V.; "Special issue: Mid-ocean ridges", Oceanus, 34(4), 5-111, 1991/92, 1992

MacLeod, C. L., P. A. Tyler, C. L. Walker (eds); "Tectonic, magmatic, hydrothermal and biological segmentation of mid-ocean ridges", Geol. Soc. Special Publication no. 118, 258 pp., Geological Society: London, 1996

Murton, B. J., L. M. Parson; "Segmentation, volcanism and deformation of oblique spreading centres: A quantitative study of the Reykjanes Ridge"; Tectonophysics, vol. 222, no. 2, p. 237-257, 1993

Parson, L.M., B. J. Murton, P. Browning (eds); "Ophiolites and their modern oceanic analogues", Geol. Soc. Special Publication no. 60, 330 pp., Geological Society: London, 1992

Phipps Morgan J., D. K. Blackman, J. M. Sinton (eds); "Mantle flow and melt generation at mid-ocean ridges", American Geophysical Union Monograph no. 71, 361 pp., American Geophysical Union: Washington DC, 1993

Smith, D.K., J. R. Cann; "Hundreds of small volcanoes on the median valley floor of the Mid-Atlantic Ridge at 24-30°N", Nature, vol. 348, no. 6297, p. 152-155, 1990

Smith, D.K., J. R. Cann; "The role of seamount volcanism in crustal construction at the Mid-Atlantic Ridge (24-30°N)", Journal of Geophysical Research, vol. 97, no. B2, p. 1645-1658, 1992

Vine, F. J., D. H. Matthews; "Magnetic anomalies over oceanic ridges", Nature, vol. 199, p. 947-949, 1963

• About hydrothermal activity:

Elderfield, H., R. A. Mills, M. D. Rudnicki; "Geochemical and thermal fluxes, high-temperature venting and diffuse flow from mid-ocean ridge hydrothermal systems: The TAG hydrothermal field, Mid-Atlantic Ridge 26°N", Geological Society Special Publication no. 76, pp.295-307, 1993

Elderfield, H., A. Schultz; "Mid-ocean ridge hydrothermal fluxes and the chemical composition of the ocean", Annual Review of Earth and Planetary Sciences, vol. 24, p. 191-224, 1996

Humphris, S. E., R. A. Zierenberg, L. S. Mullineaux, R. E. Thomson (eds); "Seafloor hydrothermal systems: physical, chemical, biological and geological interactions", American Geophysical Union Monograph no. 91, American Geophysical Union: Washington DC, 1995

Parson, L. M., C. L. Walker, D. R. Dixon (eds); "Hydrothermal vents and processes", Geological Society Special Publication no. 87, p. 257-294, 1995

Rona, P. A.; "Black smokers, massive sulphides and vent biota at the Mid-Atlantic Ridge", Nature, vol. 321, p. 33-37, 1986

Tunnicliffe, V.; "The biology of hydrothermal vents: Ecology and evolution" Oceanography and Marine Biology Annual Review, vol. 29, p. 319-407, 1991

Figure 6.1. Abyssal plains and basins throughout the world.

6

Abyssal Plains and Basins

Abyssal plains and basins are traditionally defined as areas of the deep ocean floor in which the ocean bottom is flat, with a slope of less than 1° and depths greater than 4,500 metres. They were not recognised as distinct physiographic features of the present seafloor until the late 1940s, and systematic investigations of relatively few examples have been made. As a result, they are among the least-known areas of the Earth's surface. Abyssal plains and basins, however, have an important economic potential. Transcontinental cables and pipelines are routinely laid on the bottom of deep-sea plains. Their safety and reliability relies on the correct identification of the features and geological activity observed from route surveys. During the late 1970s, an international research programme began to examine selected areas for the disposal of radioactive waste, and current studies are still looking at the abyssal plains for the disposal of environmentally harmful compounds (like for the Brent Spar oil rig in 1995). Recent studies (e.g. Müller and Holloway, 1995) have also shown the role of abyssal plains in the circulation of water masses and the modelling of ocean circulation.

6.1 GEOLOGICAL BACKGROUND

Study of the abyssal plains really started in the 1970s, with the examination of their potential for the disposal of radioactive waste. With the exception of trenches, abyssal plains often form the deepest parts of the oceans, and in the Atlantic the largest abyssal plains are located between the base of the continental slopes and the distal parts of the Mid-Atlantic Ridge, and in such topographic highs as the Madeira-Tore rise. Approximately 75 flat areas have been recognised in the world as being abyssal plains (Figure 6.1). Their monotonously flat seabed is produced by the ponding of sediments transported beyond the base of continental slopes, which causes an infilling of the deeper areas between abyssal hills. The hills become gradually submerged as the area of the plain extends, although few plains have become so mature as to have no protruding hills left. As a result of the infilling, all depressions in the seabed become flattened and there are few undulations in the seabed of more than a few metres, although small regional slopes may exist (Figure 6.2). Abyssal plains finally disappear at subduction zones. The term of "abyssal plains" is often extended to include abyssal basins. The latter are the product of tectonic and volcanic processes associated with ridge dynamics, subsequently modified by off-axis sedimentation and mass wasting. They are also associated with important depths and small topographic variations, and cover large areas of the seafloor (see Figure 6.1).

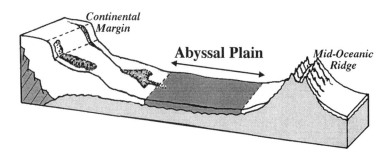

Figure 6.2. Abyssal plains and basins are the flattest areas of the seafloor, and mark the junction between mid-oceanic ridges and continental areas.

Sonar imagery of abyssal regions reveals their different styles of formation and development, such as the redistribution of sediments by local bottom currents (turbidites) and along past or present topography (contourites). Individual bathymetric features will mainly consist in abyssal hills and seamounts, most of them deeply blanketed by sediments. Seismic studies show the presence of faulting, created by mid-plate tectonic activity and differential compaction of the sediments. However, this process is not observable at the surface of the abyssal plains.

6.2 THE USE OF SUBSURFACE INFORMATION

Abyssal plains are by definition very flat and homogeneous. They are mainly filled with sediments, which look relatively similar and uniform on sidescan sonar images. Regions with different backscatter patterns can sometimes be seen, but they are less easily distinguishable and interpretable than terrains in, for example, mid-ocean ridges. Another unknown is the depth of penetration of the acoustic wave, which depends on the sonar frequency and the physical properties of the sediments (see Chapter 9: Image Anomalies and Sonar System Artefacts). This penetration effect, known as volume reverberation, is more significant at frequencies of 12 kHz or lower ($\lambda_{inc} > 12.5$ cm).

To account for this possible penetration, and to supplement the interpretation with more details, additional information is required. Sonar surveys of abyssal plains are usually coupled with seismic surveys, which provide subsurface information. But not all surveyors can afford the cost or the time needed, nor do all studies necessitate such information. We saw in Chapters 2 (Sonar Data Acquisition) and 3 (Sonar Data Processing) that most of the deep-tow sonar platforms have a depth-sounder or sub-surface acoustic profiler, which gives the height of the vehicle above the seafloor. Because it transmits at lower frequencies (typically 3.5 kHz or 7.5 kHz), the depth profiler can penetrate sedimented seafloor, and show the structure of the shallow subsurface (Figure 6.3). These profiles are acquired along-track, and are generally recorded on paper. They are represented with time (i.e. position) along-track on the horizontal axis, and travel time (i.e. distance to the transducer, or depth) of the echoes on the vertical axis.

Penetration depths vary according to the type of terrain. For example, the top profile in Figure 6.3 exhibits steep hyperbolae. Their sharp definition indicates a hard seafloor. The high variations in travel times are characteristic of bathymetric highs and lows. This corresponds to an acoustically hard seafloor (little or no sedimentation), deeply cut by gullies or troughs. As with seismic profiles, travel times exaggerate the apparent depth shown in acoustic profiler records.

The middle profile of Figure 6.3 exhibits a continuous and strong first return that gently slopes down to the right. This sharp line lies above a grey substratum in which a horizontal thick black line can be seen. This is an example of an acoustically transparent seafloor (top echo), under which are small objects with varying homogeneities or compositions. The second harder reflector beneath may be the original basement, now completely covered with softer, acoustically transparent sediments. The seafloor shows a small-scale depth variation of a few metres. In conjunction with the sidescan sonar image with which it was acquired, the data are consistent with the interpretation of a debris-flow deposit.

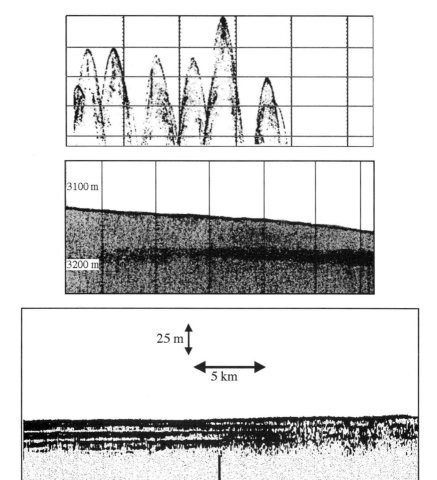

Figure 6.3. Examples of typical 3.5-kHz profiles in different terrains: (top) hyperbolae corresponding to topographic highs and lows; (middle) acoustically transparent seafloor; (bottom) transition between turbidites (left) and debris flow deposits (right).

The lower profile shows a smooth, flat seafloor underlain by two different regions (the transition is indicated by the vertical bar). Left is a transparent substrate with parallel-layered harder echoes. This corresponds to different sequences of turbidites, emplaced successively. On the right, the echoes are more complex: they change from a sharp or prolonged echo to fuzzy weakly reflective patches. These correspond to mud flow or debris flow deposits.

These are just a few simple examples of low-frequency acoustic profiler data. More systematic explanations of their relationship with sidescan sonar imagery and the local geology, calibrated with coring, dredging, and deep-tow camera surveys, can be found in Pratson and Laine (1989), O'Leary and Dobson (1992), and Masson et al. (1992).

6.3 THE FILLING OF ABYSSAL PLAINS

The abyssal plains are covered with sediments, mainly turbidites. Large-scale surveys around the world have shown that hemipelagic sediments usually make up less than 10 to 20% of the sediment input, and that debris flows are a distant third in overall importance (Pilkey, 1987). How are all these sediments transported there? The bulk comes from neighbouring continental margins (see Chapter 7: Continental Margins). It arrives via point source, i.e. submarine canyon mouths and fans. Events that furnish material from adjacent slopes without canyons tend to be small and infrequent and are only minor sources of sediments. These events are constant through long periods of time.

The first example is a GLORIA image of the Mississippi Fan (Figure 6.4). Every day, the Mississippi River discharges several millions of cubic metres of water heavily laden with sediments. The sediments are transported down the continental slope and deposited along the Mississippi Canyon and, close to the abyssal plain, on the Mississippi Fan. This sidescan image was acquired during the survey of the southern United States Exclusive Economic Zone, with the shallow-towed system GLORIA. Regularly spaced survey lines run in a SW-NE direction, and the resulting swaths have been merged into this mosaic. The terrain slopes gently down to the south-east, with water depths varying from 2,250 to 3,000 metres in the area covered by the image. The Mississippi Fan is the dominant morphologic feature of the Gulf of Mexico and appears as a large region of high backscatters extending several hundreds of kilometres. Much of the surface of the fan comprises highly reflective deposits, called depositional lobes. They all seem to have been fed from a single meandering channel system that can be traced across the fan. The largest of these is a mass-wasting deposit on the middle to upper part of the fan. The low-backscatter regions around the edges of the fan are interpreted to be fine-grained hemipelagic deposits. The surface of the fan to the west of the channel has a different acoustic texture than in the east, with radiating linear stripes of undetermined origin (but apparently related to the redistribution of sediments).

Submarine canyons are the other high providers of sediments for the abyssal plains. Connecting with the Mississippi Fan, the De Soto Channel is the highly meandering channel visible in Figure 6.5. This image was also acquired with GLORIA, and actually corresponds to a small subset of Figure 6.4. The channel shows as a bright sinuous line on a darker background, trending to the south-east. The channel is constantly less than a kilometre wide and does not show any interruption apart from the sonar tracks, where its structure is not visible. The De Soto Channel is part of an elevated channel and levee system (levee ridge). The highly reflective debris flow deposits from the Mississippi Canyon area are dammed by this elevated channel until the flow eventually overtops the levee and buries the channel. The southern extent of this channel and its deposits is masked by the debris-flow deposits of the Mississippi Fan. The bright straight line visible in the north-east corner of the image corresponds to the fault scarp of the Florida Escarpment (where the depth goes down from 2,000 m to 3,000 m in less than 3 km).

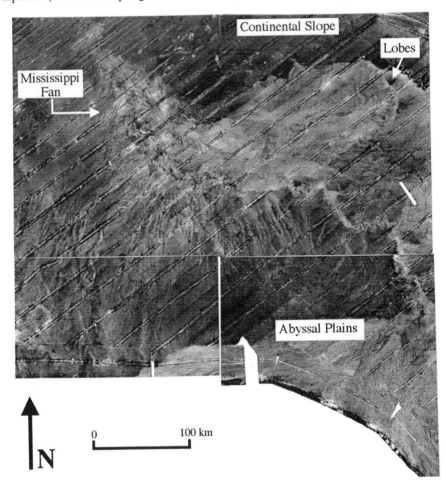

Figure 6.4. GLORIA imagery of the Mississippi Fan, Eastern Gulf of Mexico. Huge flows of sediments (turbidites) are coming from the Mississippi River. They cross the continental slope (see Chapter 7) down to the abyssal plains. Courtesy of the United States Geological Survey.

Not all processes of abyssal plain filling occur with this regularity and constancy. Recent discoveries have shown that some major abyssal plain deposits were in fact the product of huge submarine landslides (e.g. Holcomb and Searle, 1991; Masson, 1996), triggered by events up to several hundred kilometres away. One such example is the El Golfo debris avalanche, on the northern flank of Hierro Island in the Canaries (Figure 6.6). High-resolution TOBI imagery from two overlapping swaths has been

mosaicked. This image was acquired in water depths of ~ 4,000 m, while TOBI was
flying at several hundred metres above the seafloor.

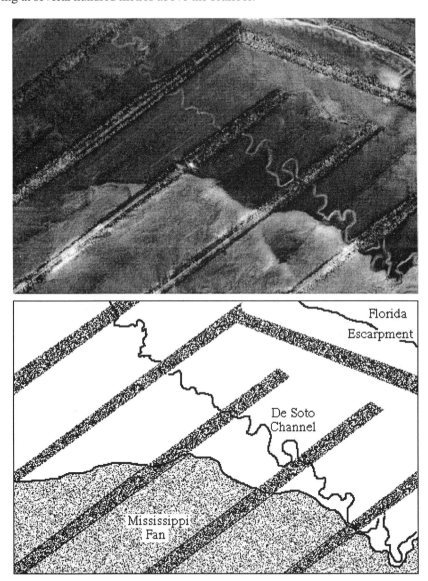

Figure 6.5. GLORIA imagery of the De Soto channel, Eastern Gulf of Mexico.
The schematic interpretation shows the nadir tracks of the survey lines (dark grey
pattern), the northern portion of the Mississippi Fan (light grey pattern) and the
highly meandering canyon in black. The image covers a ground area of 110 x 75
km. Courtesy of the United States Geological Survey.

Conspicuous large angular blocks are visible everywhere on a grey substratum composed of low-reflectivity sediments. The blocks are recognised by their higher backscatter, their rougher texture, and their jagged shapes. Some of them are up to 1.2 km across. They project important shadows which emphasize their heights and shapes (up to 200 m high). Other, smaller, blocks are in the middle of small depressions (Figure 6.7), which may have been caused by their emplacement.

The El Golfo avalanche can be traced back to a 900-m onshore fault scarp and downslope to the Madeira abyssal plain 600 km west of the Canaries. Dating of sediment cores and of the failure scarp onshore indicate an age between 13,000 and 17,000 years (Masson, 1996), therefore relatively recent. The El Golfo avalanche affected an area of 1,500 km^2, and its volume is estimated to be around 250-350 km^3. This is about 20% of the volume of the Alika landslide off Hawaii, believed to have caused the enormous tsunami that reached a height of 325 m on the neighbouring island (Moore and Moore, 1984).

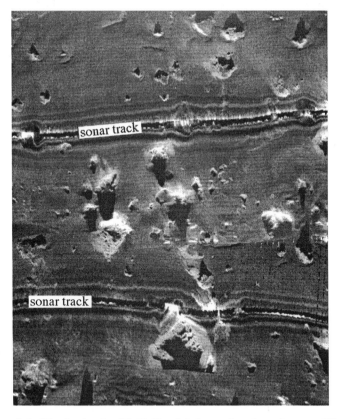

Figure 6.6. Blocks in the El Golfo debris avalanche, close to Hierro Island, Canaries. This TOBI image is made up of two overlapping swaths and is approximately 9 km high. Courtesy D. Masson, SOC (UK.)

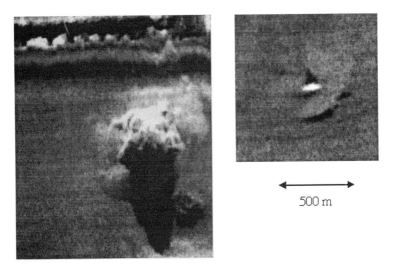

500 m

Figure 6.7. Close-up views of some blocks in the avalanche debris shown in the previous figure. They project long shadows which emphasise their heights. Their faces facing the sonar are very bright, which means that they have steep slopes. Courtesy D. Masson, SOC (UK.)

6.4 STRUCTURES OF THE ABYSSAL PLAINS

In abyssal plains, the pre-existing bathymetry plays an important role in determining flow paths and turbidite thicknesses. The individual turbidity currents produce tongue-shaped turbidites elongated in the direction of the flow. They are visible on sidescan sonar images because of a higher backscatter and a texture more mottled than their surroundings (Figure 6.8).

The production of turbidites with other shapes is usually due to material arriving on the plain from several points at the same time, combining to form a single turbidite system, or to the restriction of the flow by topographic contours (the turbidites are then named contourites). One of the best studied abyssal basins in the world is the Canary Basin, west of the Canary Islands and at depths generally close to 5,000 m. The Canary Basin is a dynamic area with periodic erosion and transport of hundreds of cubic kilometres of sediment, accomplished through debris flows and turbidity currents. The next examples are all located in this area, which has been extensively studied by our colleagues at the Southampton Oceanography Centre (see the General Bibliography).

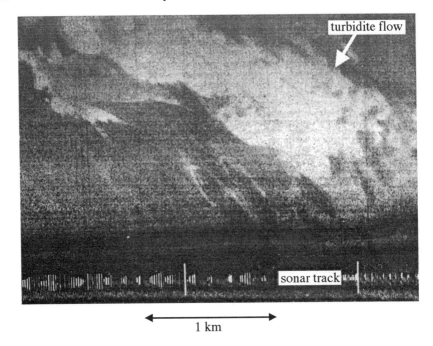

Figure 6.8. Raw TOBI image of relatively fresh turbidites (lighter tones) in the Madeira Abyssal Plain. Because of the absence of correction, backscatter levels are attenuated with distance from the sonar track. Courtesy D. Masson, SOC (UK.)

Figure 6.9 shows an example of a debris slide overlying an older channel. The channel is the large grey area crossing diagonally the image. Its texture is smoother and more organised. The overlapping slide is brighter, and its fabric less well-developed, suggesting a rougher scree-like surface. The lobate edges of the slide are interpreted as small slumps from the new deposits. They can also be attributed to turbulent conditions at the time of their emplacement.

Slumps occur when sediments accumulated on a slope are undermined by subsurface movements and collapse. Figure 6.10 shows a well-defined slump imaged with TOBI in the Madeira Abyssal Plain. The slump was ensonified from the right of the image. The arcuate pattern, made up of small concentric ridges and troughs, is characteristic. It bends in the direction in which the slump occurred, i.e. to the right in this case. The sediments in the rightmost part of the image are undisturbed, whereas those in the left parts show textures more and more disturbed as they are close to the main deformation area. Slumps are not restricted to abyssal plains and basins, and can occur in all areas with sufficient sediment instability (see Chapters 7 and 8).

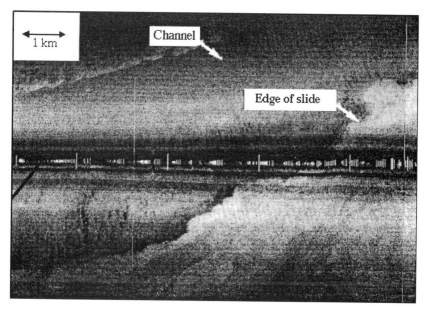

Figure 6.9. Uncorrected TOBI imagery of a debris slide superimposed on an older turbidity channel. Courtesy D. Masson, SOC (UK.)

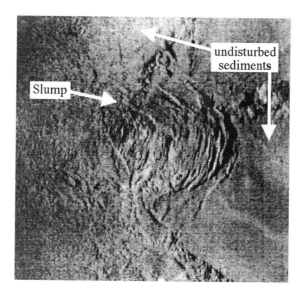

Figure 6.10. TOBI imagery of a slump in the Madeira Abyssal Plain. Ensonification direction is from the right, and the horizontal dimension is approximately 1 km. Courtesy D. Masson, SOC (UK.)

The surface of debris flows will appear differently according to the frequency and resolution of the imaging sonar. Surfaces which looked uniformly dark and featureless with GLORIA exhibit with TOBI grainy textures organised along preferential directions. Bands of differing backscatter intensity correspond to fine-scale flow banding along the main directions of sediment transport. If they are nearly parallel to the sonar track, these lineations are more difficult to see (Figure 6.11) than if there is a substantial angle (Figure 6.12). This may be due to the small distances between the different lineations. The distance between two successive lineations will appear larger if it is imaged in an oblique line, and therefore will stand more chances of being visible on the sonar record. Flow banding is on a scale of tens to hundreds of metres. Sidescan images show no indication of relief, except rare tiny acoustic shadows at the edges of the bands. This indicates the relief is small, probably less than 1 metre. Flow banding is considered to result from flow streaming of clasts, with variations in clast size between bands affecting the amount of backscatter.

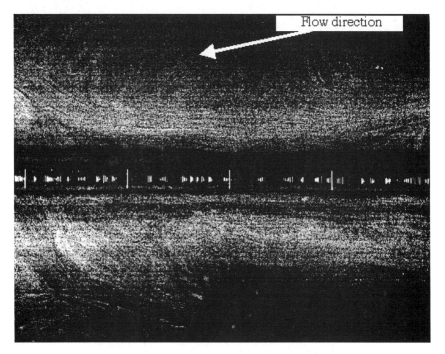

Figure 6.11. TOBI imagery of flow-banding sub-parallel to the sonar. This image and Figure 6.9 are at the same scale. Courtesy D. Masson, SOC (UK.)

Some of the flow lineations are oriented with varying angles, showing buckles reminiscent of structures under compression. These features are pressure ridges. They have been observed in some sidescan images of other submarine sediment slides, and are usually seen in subaerial landslides, mud slides and sedimentary facies (e.g. Price and Cosgrove, 1990). The orientation of the indentations and their wavelengths give details about the direction and strength of the compression they have undergone.

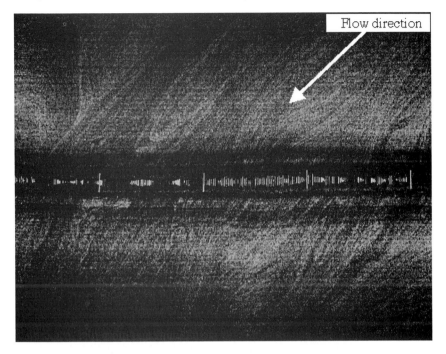

Figure 6.12. TOBI imagery of flow-banding oriented obliquely to the sonar.
Courtesy D. Masson, SOC (UK.)

Figure 6.13. Pressure ridges as seen on TOBI imagery. The scale is the same
as in Figure 6.12. Courtesy D. Masson, SOC (UK.)

This primary fabric is cut by a series of distinct flow-parallel longitudinal shears. The cross-cutting fabric visible on the sonar imagery consists of distinct lines, often truncating the fine-scale banding and separating areas of different fine-scale trends. The lines of the cross-cutting pattern are hundreds of metres to kilometres apart, and most are subparallel to the regional slope. In one area, one such lineament appears to be deformed (Figure 6.14). Many of these lines cast narrow but distinct acoustic shadows, and this relief must be about 1 m or less. It appears that this element of the "woodgrain facies" represents a series of longitudinal shear zones marking differential movement between streams of debris which moved at different speeds or possibly at different times.

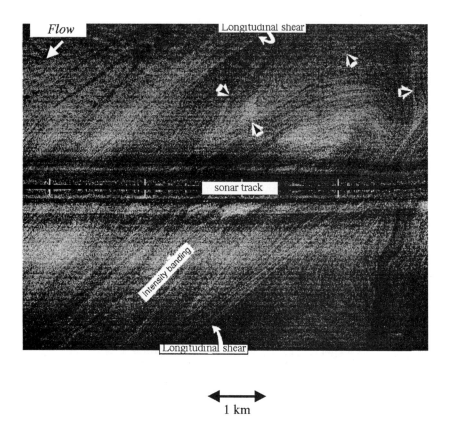

Figure 6.14. Longitudinal shear structures on TOBI imagery. The small arrows point out an imbricate structure, barely visible in the image. Courtesy D. Masson, SOC (UK.)

TOBI imagery from the abyssal plains sometimes show subparallel lineations with a high backscatter (Figure 6.15). They are 10-50 m wide, spaced 20-200 m apart, and are in this case aligned along the slope. Some individual lineations can be followed across the full extent of the swath, meaning they are at least 6 km long. They are usually interpreted as erosional grooves. This is shown here by their geological context. The lineations are at the limit between a debris flow (highly textured region in the bottom left corner of the image) and undisturbed sediments (smooth darker region in the top right corner of the image). They are overlain by the edge of the debris flow and aligned downslope. All these arguments concur with an erosional origin for these features. This acoustic character of erosional features is also visible in other types of environments (see Chapter 8: Coastal Environments).

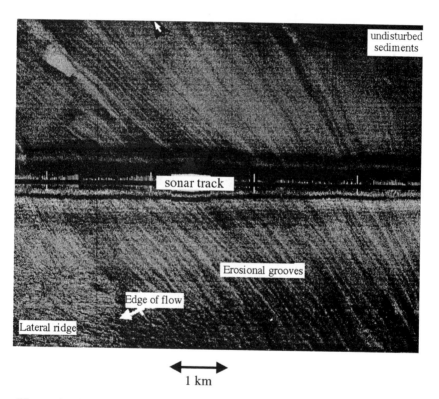

Figure 6.15. Erosional structures as seen with TOBI imagery. The small arrow in the top of the image indicates the downslope direction. Courtesy D. Masson, SOC (UK.)

6.5 RELICT STRUCTURES

The high input of sediments in the abyssal plains does not always obliterate pre-existing structures. These relict structures may come from previous sediment flows, or from the original basement on which the sediments have accumulated. In the Madeira Abyssal Plain, large pieces of oceanic crust are still visible. They extend for several tens of kilometres (Figure 6.16) and have been dated to 100 million years BP (before present). The sidescan sonar image shown in Figure 6.16 was acquired with TOBI. It shows part of an arcuate ridge of old material, darker than its surroundings. It lies on top of undisturbed sediments, recognisable by their smooth homogeneous textures (lower portion of the image). The alternating bright scarps and their associated acoustic shadows suggest the older material is higher than the surrounding seafloor. The large dark block in the middle of the image is approximately 2.5 km wide, and shows a smooth flat top. Cores taken on the top showed it to be an old seamount, with a shallow sedimentary cover which is easily penetrated by the acoustic energy from the sonar. The dark region behind the seamount also belongs to the piece of underlying oceanic crust.

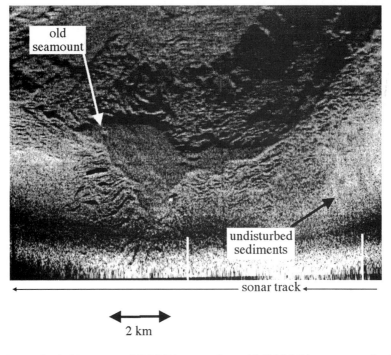

Figure 6.16. Uncorrected TOBI image of an old (100 Ma) seamount in the Madeira Abyssal Plain. Courtesy D. Masson, SOC (UK.)

Smaller blocks with low backscatters are visible in other portions of the abyssal plains. Their dimensions may range from a few hundreds of metres to a few kilometres, and their shapes are rounded to irregular. They are usually associated with bright haloes of coarsely textured material. One example of these blocks is presented in Figure 6.17. The relict block is recognisable by its very dark uniform backscatter. It is surrounded by streamlined haloes of coarsely textured rubble, forming a distinct teardrop shape. The teardrops are elongated down the slope, and aligned along the apparent flow direction. The streamlined structures may be formed by material aggregated to the edges of the block, or by the progressive fragmentation of the block's edges in the flow. This particular block is a relict block, part of an older structure now largely buried under sediments. Alternatively, it could be interpreted as a rafted block, i.e. a coherent sediment block rafted by the flow to its present position. The distinction between the two hypotheses would be made on the dimensions of the block (it should not be much thicker than the flow deposit which transported it) (50 m in the present case), and if possible on additional ground-truthing data such as cores.

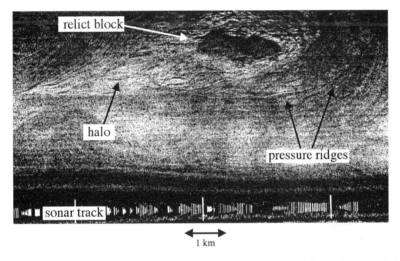

Figure 6.17. TOBI imagery of a relict block in the Madeira Abyssal Plain. Courtesy D. Masson, SOC (UK.)

Similar structures can be witnessed in coastal waters (see Chapter 8). These configurations are morphologically similar to their terrestrial equivalents. The sediments around islands in a rapidly flowing river exhibit a similar shape. So do the yardangs, positive features shaped like inverted boat hulls and formed by high wind erosion of old lava flows. Spectacular analogues to these features are also observed on Mars. The planet is thought to have been at one time submitted to catastrophic flooding, huge rivers carving out large canyons such as Valles Marineris and producing important flows. These flows were deflected by pre-existing structures and created similar teardrop shapes (Figure 6.18).

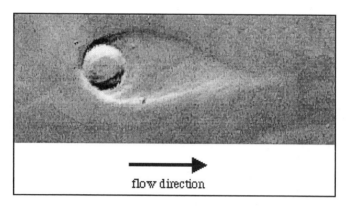

flow direction

Figure 6.18. Martian analogue of a relict block and a teardrop-shaped deflection of the flows. In this case, the block is an 8-km wide impact crater formed by a meteorite. The flows were created by the intense floods thought to have prevailed on Mars a long time ago. They created a moat about 600 m high. The scale of these features is much higher than on Earth, but their size/height ratio and their morphology are similar to those of terrestrial features. This image was taken in the optical wavelengths by the NASA Viking satellite. Copyright NASA-JPL.

6.6 EXAMPLE OF REGIONAL IMAGERY

There are not many examples of regional imagery from the abyssal plains. Mainly because the abyssal plains have not been much explored with sidescan sonar, the surveys are usually limited to specific zones, or to long swaths covering one large geological object (e.g. Saharan debris flow). The image presented in this section (Figure 6.19) was acquired by GLORIA during the survey of the US Exclusive Economic Zone around the Hawaiian Islands. It covers the abyssal plains themselves, and the whole slope between the islands of Oahu and Molokai along with their debris avalanches.

The image is a mosaic of several GLORIA swaths oriented SE-NW, and a few additional swaths around the islands of Oahu and Molokai in the southern part of the image. The water depths are rapidly varying from about 1,000 m to 4,500 m in the main part of the slope (southernmost portion of the image), and are then stabilised between 4,500 m and 5,000 m in the rest of the image. These important variations yield large differences in swath widths between the lines close to the islands, and the lines further out in the abyssal plain. Each swath was fully processed for radiometric and geometric corrections.

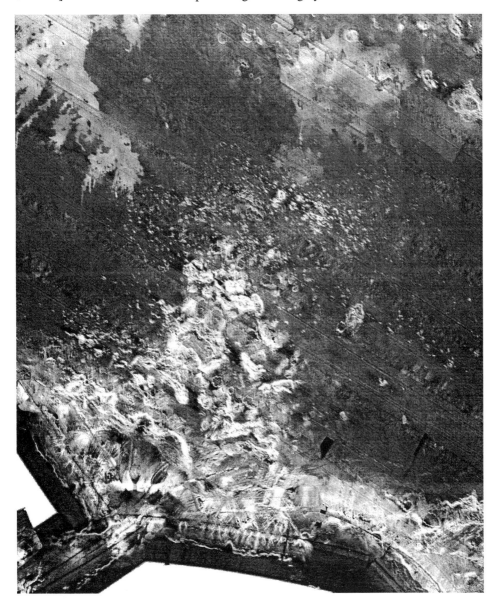

Figure 6.19 GLORIA mosaic of region north of the Hawaiian islands of Oahu and Molokai, showing the important debris avalanches that spilt from the shores down the abyssal plains. The whole image covers an area of 228 km by 274 km, and is oriented with the North on top. Courtesy D. Masson, SOC (UK.)

The image can be readily divided into 5 major parts. The first one, in the southernmost region and close to the islands, exhibits a very bright texture, organised perpendicularly to the islands. The second unit is made up of large reflective blocks on a mottled substrate, and points out like a tongue toward the middle of the image. The third and largest unit visible is dark and featureless (apart from the superposed blocks at the limits of the previous unit). The fourth and fifth units are smaller and well-delimited medium to highly reflective patches.Close-up views of these different zones further illustrate their differences (Figure 6.20). In the far north of the image, two very distinctive regions are seen. They have smooth irregular boundaries, and exhibit medium-high to high backscatters and very fine-grained textures, but are nearly homogeneous. No relief is visible, either from variations in backscatter or from clear acoustic shadows. Because its appearance resembles lava flows (see Chapter 5: Mid-Ocean Ridge Environments), albeit larger than any seen at mid-ocean ridges, this region is interpreted as shallow buried lava flows (Figure 6.21). A thin layer of sediments is visible on profiler records, but is transparent to GLORIA's acoustic waves (GLORIA's penetration depth is related to its relatively low frequency of 6.5 kHz). The same type of region is visible in the north-east corner of the regional image: same shapes, same texture, and slightly less reflective. This difference is attributed to a thicker sediment layer above the lava flows: the backscattered waves are more attenuated. This is confirmed by profiler records along the different survey lines. A few rounded blocks are visible in the buried lava flows (Figure 6.20), with a higher backscatter on their slopes facing the sonar. They have dimensions of a few kilometres and belong to the Hawaiian Arch a few tens of kilometres to the north.

Most of the image is made up of a dark, featureless substratum. These are the abyssal plain sediments (Figure 6.20). Around the edges of the tongue-shaped region, they are littered with small debris blocks, with dimensions of a few hundred metres, some of them at the limit of resolution of GLORIA. Larger debris blocks are seen further upslope, closer to the two islands (Figure 6.19). These blocks have dimensions of several kilometres, up to 50 km^2 for the largest one. They are highly reflective, with rounded to elongate shapes, and they extend 100 km from the fault scarps at the base of the islands' slopes (Figure 6.20). The blocks are very large at the beginning (some of them have volumes of several cubic kilometres) and then diminish in size with distance from the origins (much like terrestrial landslides). Only the smallest blocks litter the sediments on the seafloor, more than 200 km away from their sources on the islands.

Amidst these large blocks, an elliptical patch of darker reflectivity can be noticed (Figure 6.20). It is bounded by highly reflective contours, corresponding to sonar-facing fault scarps, and acoustic shadows in the NE where the slopes are facing away from the sonar. The darker patch itself is quite flat and homogeneous, and forms a large plateau on top of the structure. The whole object seems to have blocked the pathway for the larger blocks coming from the west (Nuuanu Slide). This structure therefore predated the avalanche slide. Because of its size and geological setting, it is interpreted as a relict volcano (its name is Tuscaloosa Seamount). Other distinct structures include the sediment channels just north of Molokai Island (Figures 6.20 and 6.21). Sediments and debris coming from onshore locations are following the topography and form gullies and canyons. The shapes of these structures are similar to the submarine canyons found on continental margins and shown in the next chapter (Figures 7.4 and 7.5).

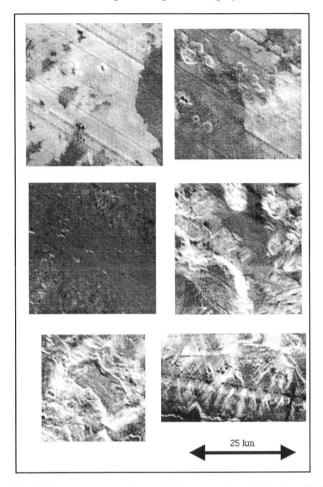

Figure 6.20. Close-up views of the different geological units visible in the regional image of Figure 6.19. From left to right and top to bottom: shallow buried lava flows; more deeply buried lava flows; abyssal plain sediments with small debris blocks; fault scarps at the base of the fault slope and large debris blocks; the flat-topped Tuscaloosa Seamount, and sediment-channelling structures north of Molokai.

These distinct elements build up a story that can be unfolded from the islands' slopes downward. First, several submarine canyons feed sediment down the slope, where it has accumulated to produce a 500-m thick deposit (Moore et al., 1989). The NW canyons are clearly the continuation of subaerial canyons, and are thought to have been carved on shore and drowned during the subsidence of the Hawaiian Ridge under the weight of the growing volcanoes. Because the canyon morphology is well developed down to 1,800 m depths, this value is presumed as the amount of subsidence undergone.

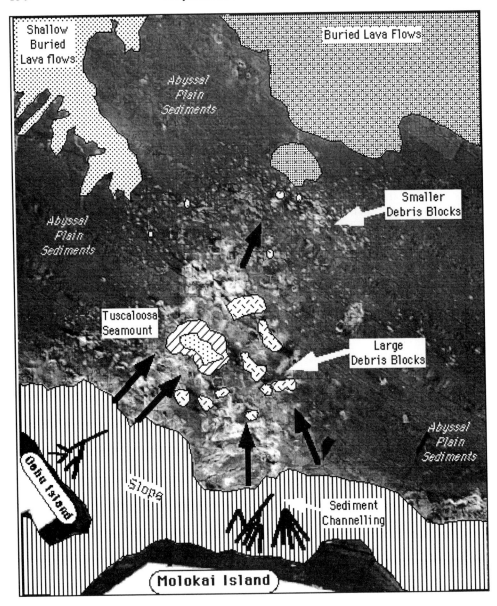

Figure 6.21. Schematic interpretation of the regional GLORIA imagery. The black arrows indicate the directions of the two avalanche slides, Nuuanu in the west and Wailau in the east.

The two catastrophic avalanche slides are the Nuuanu Slide in the west and the Wailau slide in the east. Lying to the north of Oahu Island, the Nuuanu Slide is the largest mapped along the Hawaiian Ridge. It was formed by the collapse of the northern flank of the Koolau volcano along a volcanic rift zone. Onshore, the line of failure is marked by a line of steep cliffs, the Nuuanu Pali, that are parallel to the coastline. The Nuuanu debris avalanche deposit forms an amalgam with the adjacent Wailau debris avalanche deposit, a similar slope failure deposit originating from the northern flank of the neighbouring island of Molokai. Together, they cover around 25,000 km² of the seafloor.

The debris avalanche deposit extends for about 230 km from its head-wall at the Nuuanu Pali to its toe. It crosses and partially in-fills the large bathymetric depression known as the Hawaiian Deep, then runs upslope for about 120 km, with a vertical rise of around 300 m (Moore et al., 1989; Jacobs, 1995). Geophysical models, coupled with seismic surveys, make a conservative estimate of a volume of about 5,000 km³ for the total avalanche, ranking it as one of the largest mass-movement deposit of any type discovered so far on Earth. These slope failures apparently began early in the history of the individual volcanoes, when they were small submarine seamounts. They culminated near the end of subaerial shield building and continued long after dormancy. Their importance and possible recurrence emphasize the need for a precise charting of the seafloor around major volcanoes, which should help assess the risks associated with massive landslides and the associated tidal waves.

6.7 SUMMARY

The abyssal plains are the flattest and most extensive regions of the seafloor. They exhibit nearly featureless surfaces and have therefore not been studied extensively with sidescan sonar. We saw in this chapter the main structures that were associated with the filling of abyssal plains (sediment channels, debris slides and flows), and those associated with redistribution of sediments in the plains (turbidite flows, flow-banding and relict structures). A better knowledge of these redistribution patterns is essential for the sometimes envisaged deep-sea disposal of toxic materials. There is no doubt that further sidescan sonar surveys of the abyssal plains will concentrate on these aspects. It is only very recently that high-resolution sonar imagery has revealed the surface expressions of flow-banding and shear in mass-sediment flows (Masson, 1996) and new insights are being gained from their detailed information.

The avalanche slides from the Hawaiian volcanoes of Oahu and Molokai are particularly spectacular examples. These huge structures were recently discovered, as were the evidences that they created major tsunamis at the time. The human and economic implications mean that similar structures are likely to be extensively mapped in the near future, and their potential risks assessed.

A subject we did not touch upon is the anthropogenic influence on the abyssal plains. Its manifestations can be summed up as intentional man-made structures (cable lines and pipelines), and unintentional traces (shipwrecks, trawl marks). They are occasional and restricted in space. Some images of these manifestations are shown in the next chapters: Chapter 9 for submarine cables, and Chapter 8 for shipwrecks and trawl marks.

6.8 FURTHER READING

*   **About abyssal plains in general:**

Kidd, R.B., R.W. Simm, R.C. Searle; "Sonar acoustic facies and sediment distribution on an area of the deep ocean floor"; Marine and Petroleum Geology, vol. 2, no. 3, p. 210-221, 1985

Weaver, P.P.E., J. Thomson; "Geology and geochemistry of abyssal plains", 246 pp., Blackwells for the Geological Society,. Oxford, 1987

*   **About sonar imagery and subsurface information**

Masson, D.G., R.B. Kidd, J.V. Gardner, Q.J. Huggett, P.P.E. Weaver; "Saharan continental rise: facies distribution and sediment slides", p. 327-343. In Poag, C.W., P.C. de Graciansky; *Geologic Evolution of Atlantic Continental Rises*, 378 pp., Van Nostrand Reinhold, New York, 1992

Pratson, L.F., E.P. Laine; "The relative importance of gravity-induced versus current-controlled sedimentation during the Quaternary along the middle U.S. continental margin revealed by 3.5-kHz echo character", Marine Geology, vol. 89, p. 87-126, 1989

*   **About submarine landslides:**

Holcomb, R.T., R.C. Searle; "Large landslides from oceanic volcanoes", Marine Geotechnology, vol. 10, p. 19-32, 1991

Jacobs, C.; "Mass-wasting along the Hawaiian Ridge: Giant debris avalanches", p. 26-28. In *Atlas of Deep Water Environments: Architectural style in turbidite systems*, K.T. Pickering, R.N. Hiscott, N.H. Kenyon, F. Ricci Lucchi, R.D.A. Smith (eds.), 333 p., Chapman & Hall, London, 1995

Masson, D.G.; "Catastrophic collapse of the volcanic island of Hierro 15 ka ago and the history of landslides in the Canary Islands", Geology, vol. 24, no. 3, p. 231-234, 1996

*   **About deep-sea debris flows:**

Masson, D.G., Q.J. Huggett, P.P.E. Weaver, D. Brunsden, R.B. Kidd; "The Saharan and Canary debris flows offshore Northwest Africa", Landslide News, Tokyo, no. 6, p. 9-13, 1992

Masson, D.G., Q.J. Huggett, D. Brunsden; "The surface texture of the Saharan Debris Flow deposit and some speculations on submarine debris flow processes", Sedimentology, vol. 40, p. 583-598, 1993

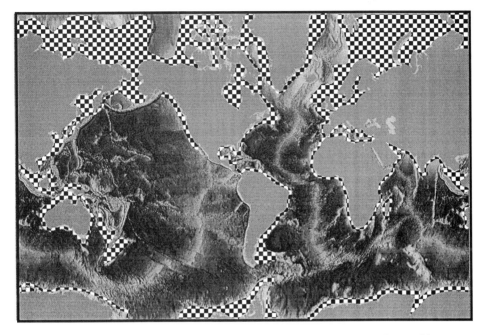

Figure 7.1. Continental margins and marginal seas throughout the world.

7

Continental Margins

In Chapter 4, the brief introduction to the different geological and physiographic provinces showed the great importance of continental margins (Figure 7.1). Although less varied geologically than mid-ocean ridges, they influence many aspects of our socio-economic activities and of the world's ecology. The Exclusive Economic Zones of most countries are occupied by continental margins, from which they derive an important portion of their wealth. Hydrocarbon exploitation is concentrated on continental shelves, for example in the North Sea or the Gulf of Mexico. The world's largest fish reserves are located in the continental slopes and shelves, and their management often proves politically and economically difficult. Slope failures and the resultant flows are capable of the destruction of marine installations and submarine telecommunications cables, and even of generating tidal waves (tsunamis). Recent environmental studies have also shown the importance of a more complete knowledge of continental margins with the linking of coastal pollution and the off-shore dumping of harmful chemical products in poorly known areas that were thought deep and stable enough.

7.1 GEOLOGICAL BACKGROUND

For all the reasons outlined in the previous section, and also because of their higher accessibility, continental margins have become the best and most studied areas on the seafloor. They can be divided into three distinct regions, outlined in Figure 7.2. Closest to shore, the continental shelf is a relatively flat area (slopes less than 1:1,000, low local relief). Supported by continental crust, it is also remarkably shallow; generally less than 250 m. Depending on the regional geological setting, the shelf will extend from a few kilometres (near subduction zones, for example) to several hundreds of kilometres (near passive margins) away from the shores. The continental shelves are submitted to the constant input of large quantities of sediments by the rivers, sometimes at large distances from land (e.g. the Indus or the Amazon). The shelves are also affected by pelagic rain.

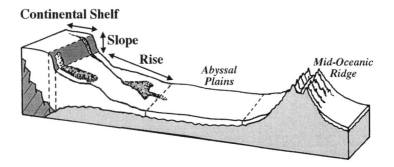

Figure 7.2. Schematic representation (not to scale) of the major physiographic province, showing the relatively flat continental shelf and the rapidly descending areas of the continental slope and rise.

These large volumes of sediments follow the bathymetry gradients down the continental slope (Figure 7.2). This rapidly descending area ranges in depth between approximately 200 m and 2,000 m. The sediments are transported along channels and canyons, far out into the abyssal regions. They also conglomerate to form mounds, structures following the bathymetric contours (contourites), etc. Because of the important slopes, some of the structures may be unstable, and earthquakes or minor sea-level changes can trigger their collapse, creating landslides (the larger ones producing tidal waves).

The sediments are carried over, down to the continental rise (Figure 7.2). This physiographic province marks the limit between the continental margins and the abyssal plains (see Chapter 6: Abyssal Plains and Basins). Deeper than 2,000 m, the continental rise can go down to 6,000 m (such as in the North Atlantic). The more prominent structures visible are the distal fans marking the limits of the canyons and sediment channels.

For the sake of simplification, and because they exhibit similar surface structures, we have included in the present chapter the epicontinental marginal seas, such as the North Sea, the Gulf of Maine or the Gulf of St. Lawrence, and the marginal plateaus. Morphologically similar to continental shelves, the marginal plateaus lie at slightly greater depths and are separated from the shelves by incipient continental slopes. The Blake Plateau in the North Atlantic is a good example. Other marginal plateaus can be found off the coasts of southern Argentina and eastern New Zealand.

7.2 SEDIMENTARY STRUCTURES

7.2.1 Sediment Deposition and Erosion

Continental margins are submitted to the continuous input of sediments from the continent and of biogenic carbonate deposits. They are deposited in large homogeneous patches, well-stratified and fine-grained, which offer a generally smooth and poorly reflective acoustic appearance. When accumulated on dipping units, unconsolidated sediments may move downslope as distinctly bounded masses ("mass movement"). These sediments may also be transported by turbidity currents which act as important erosion factors. The interplay of depositional, erosional, and mass movement processes creates complex sedimentary patterns, reflected in the variations of backscattering characteristics in the sonar imagery.

Figure 7.3 shows a perfect example of these different processes. It was collected with GLORIA on the Eastern USA continental margin, during the survey of the American Exclusive Economic Zone (EEZ-SCAN, 1991). It was acquired in water depths of 750 metres, above the Blake Plateau. The image is a mosaic of several adjacent GLORIA swaths, with a ground resolution of 50 m approximately. The sidescan sonar imagery was radiometrically and geometrically corrected with the USGS version of WHIPS (see Chapter 3). The image is clearly divided into two types of regions: one smooth and very dark, the other brighter and with a mottled texture. In the lower portion of the image, the dark region exhibits a very homogeneous texture, with no visible feature whatsoever. It appears completely flat and uniform, and is interpreted as a patch of sediments (ground-truthing reveals it to be carbonate mud). Conversely, the top portion of the image shows areas of heterogeneous bright backscatter. Some areas are randomly textured, and are outlining blocks. They are interpreted as a rock outcrop. The darker linear narrow channels that cut it at regular intervals and along the same diagonal direction are interpreted as scours created by the turbidity currents and filled by the incoming sediments. The backscatter variations over the outcrop itself are related to its micro-scale roughness, i.e. the composition (or induration) of the bedrock and, in this particular case, the presence of corals (seen from dredging and bottom photographs).

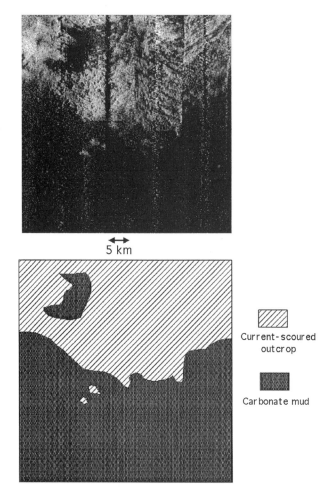

5 km

Current-scoured outcrop

Carbonate mud

Figure 7.3. GLORIA imagery of the boundary between a patch of carbonate mud and a current-scoured rock outcrop on the Blake Plateau (Eastern USA continental margin) at a mean depth of 750 metres. Courtesy United States Geological Survey.

7.2.2 Sediment Transport - Submarine Canyons

A large portion of the accumulated sediment moves downslope, driven by gravity and bottom currents. They are transported through large slides, often called debris slides, or along narrow channels merging into submarine canyons. The processes leading to either one of these two modes are complex and not entirely understood (see Hagen et al., 1994). The major controlling factors are the type of turbidity currents, their importance and velocities, as well as the depth gradients and the geology of the local substratum.

In many areas, the continental slope is cut by canyons. Most of them start on the continental shelf, and act as channels for the transport of sediment down to the abyssal plains. Submarine canyons are created by the erosive action of the sediments suspended in the turbidity currents. They are usually V-shaped and their morphology is reminiscent of river valleys on land. Very large areas extending hundreds of kilometres off-shore are covered by different types of sediment flows. They are more reflective acoustically than deposited sediment patches because of their higher micro-scale roughness and their large-scale hummocky topography. The local textural variations are related to the different processes that led to their creation and evolution. A good example of these processes is presented in Figure 7.4. This GLORIA image was collected during the survey of the US Exclusive Economic Zone (EEZ-SCAN, 1991), and covers approximately 200 by 200 km. It regroups several swaths with a ground-resolution of 50 metres. The water depth varies between -250 metres in the upper-left portion of the image, and -3,000 metres in the lower-right portion. This explains the differences in the swath widths which increase with depth.

On the basis of the textural patterns, the image can be readily decomposed into three large units: the continental shelf (top-left) with a dark, nearly-homogeneous reflectivity; the continental slope with a medium reflectivity and organised textures going downslope; and the continental rise in the lower-right with lighter reflectivities and more random textures. In the continental slope, long ribbons of high backscatter can be seen, nearly parallel to each other. Close-up views show they are closely associated with dark shadows and therefore have a lower relief than their surroundings. These channels often merge with other channels, and pour sediments into larger canyons such as the Wilmington and Baltimore Canyons. It is not rare to see canyons extending for tens or even hundreds of kilometres, depending on the local slopes. Their sinuosity will vary according to the bathymetric gradients, the flow of sediments and the local geology. Large patches of brighter reflectivity are also visible in Figure 7.4 at the base of the continental slope: these are debris flows, shown in better detail in section 7.2.3.

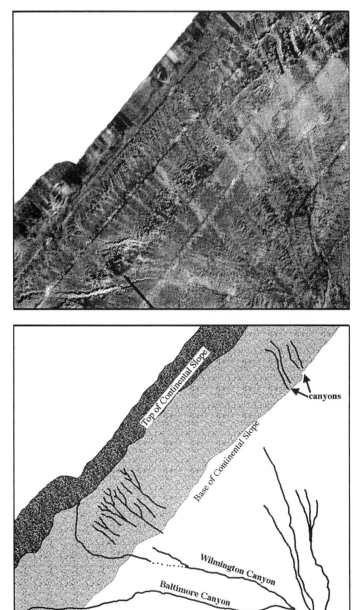

Figure 7.4. Sediments are transported down the continental slope through narrow channels, which merge into canyons. These structures often span tens to hundreds of kilometres. GLORIA imagery, courtesy United States Geological Survey.

A closer view at the boundary between the continental shelf and the top of the continental slope shows the dendritic channels along which sediments are transported (Figure 7.5). The continental shelf (upper left corner) is covered with homogeneously textured sediments. A treillis-like drainage pattern marks the limit of the slope. In a few hundreds of metres, it organises itself into cone-like patterns of bright reflectivity, which merge and will later feed the main canyons (in this case, Wilmington Canyon).

Figure 7.5. Detail of Figure 7.4, showing the dendritic arrangement of sedimentary channels. The image is 35 km wide by 35 km long GLORIA imagery, courtesy United States Geological Survey.

Not all submarine canyons look alike. Some die out in the continental slope and some extend far into the continental rise. Some are large, others quite narrow. For example, Baltimore and Wilmington canyons, presented in Figure 7.4, were fed by several channels and in some places very close morphologically to their surroundings. Conversely, the neighbouring Veatch and Hydrographer Canyons extend for nearly 200 km across the Western Atlantic continental rise as tightly sinuous channels with no tributaries (Figure 7.6). The Hydrographer Canyon is even two times narrower than Veatch Canyon. All canyons are conduits for sediments eroded from the slope, but the extents of transport and the modes of delivery have varied, and produced these different morphologies.

The GLORIA imagery presented in Figure 7.6 was acquired during the same survey of the North-American EEZ. Water depths vary from 250 metres on the continental shelf (Figure 7.6) to 4,250 metres at the southern extremity of Veatch Canyon. The image is

approximately 300 km long by 200 wide. It results from the mosaicking of many
swaths with varying orientations: parallel to the slope in the shallower parts closer to
the shelf, and wider swaths going upslope and downslope in the deeper regions.
Several geological units are easily distinguished from their backscatters and the
orientation of their textures. They correspond to sediment slides oriented in the
directions of the slopes: Veatch Slide is brighter and has a finer-grained texture than
Nantucket Slide. A larger unit of dark backscatter is visible in the lower-left portion of
the image, with a coarser texture, interpreted as mass-wasting deposits. But the main
distinctive features are the two elongated structures crossing diagonally most of the
image. They are delimited by parallel bright and dark lines, corresponding to the walls
respectively facing toward and away from the imaging sonar (Figure 7.7). These
structures are lower than their surroundings and narrow, which along with their context
identifies them as canyons. A long linear fault is visible as a dark feature crossing three
of the lower-left swaths.

Figure 7.6(a). GLORIA imagery of Veatch Canyon. Courtesy United States
Geological Survey.

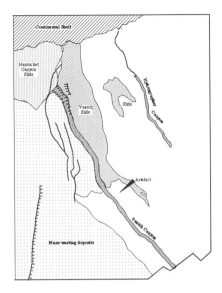

Figure 7.6(b). Geological interpretation. The two main canyons (Veatch and Hydrographer) are fed with sediments coming from the continental shelf and slope. Very narrow, these canyons extend for several tens of kilometres down to the continental rise.

Figure 7.7. Close-up view of Veatch Canyon. GLORIA imagery, courtesy United States Geological Survey.

A close-up view of Veatch Canyon shows in greater detail the walls of the V-shaped canyon, as well as the mostly homogeneous surroundings (Figure 7.7). A large sonar artefact is visible in the middle of the image showing the shape of the sonar beam. The first targets apparently correspond to Veatch Canyon itself, which would argue for a problem in the detection of the first return for this particular beam (see Chapter 9).

In particular geological contexts, and when bathymetric gradients in the continental slope are relatively small, submarine canyons can form meanders. These meanders are comparable to those of land-based rivers such as the Mississippi or the Seine. Meanders are believed to develop as a stream attempts to maintain the optimal channel slope for its sediment load on varying valley slopes (e.g. Hagen et al., 1994). Sinuosity in submarine fan channels have been observed on steeper slopes than on land, which may be due to the smaller density difference of turbidity current vs. water compared to water vs. air. Observations and experimental studies show that sinuosity in terrestrial rivers increases with the slope until a threshold is reached, at which point a relatively straight braided channel develops. The abandoned channel segments are called cut-off meanders. Cut-off meanders may also be created by the occasional input of large rapid turbidity currents, which would have difficulty negotiating the tight canyon meanders and would expend their energy against the canyon walls (much as springtime rivers coming from glaciers would "jump" over the meanders in a plain and create new river paths).

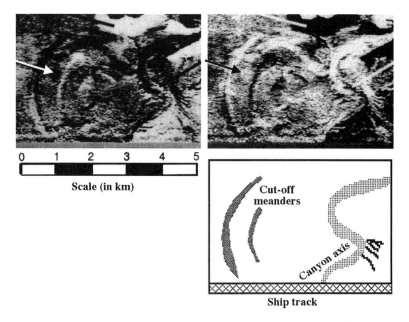

Figure 7.8. (Left) a highly-meandering canyon and two cut-off meanders imaged with SeaMARC II ; (right) the same image, inverted to show grey levels proportional to the backscatter values; (bottom) schematic interpretation. Courtesy Rick Hagen (AWI, Germany) and Elsevier Publications.

The image presented in Figure 7.8 was acquired with SeaMARC II on the continental margin off Peru (Hagen et al., 1994). SeaMARC II images are traditionally represented with lower backscatters brighter. To be consistent with the representation scheme used in this book and with the conventions of remote sensing (see Chapter 3: Sonar Data Processing), this image has been inverted in order to represent the highest backscatter values as white and the lowest backscatter values as black (Figure 7.8, right). This particular image corresponds to only one half of the swath, and therefore a constant illumination direction. The canyon axis is clearly visible on the right with its meanders, with the walls facing the sonar being highly reflective. Cut-off meanders on the terraces surrounding the canyon are clearly recognised by the sharp-edged reflections and shadows from the former channel walls, and the high areas in the centre of the meander loops. Along the southern part of the current canyon axis, small channels perpendicular to the levees are visible. They correspond to crevasse splays where canyon overflow became focused.

7.2.3 Mass-Wasting, Slides and Flows

The continental rise is formed by the processes that transport sediments in a downslope direction. Large sediment accumulations may form at the edge of the continental shelf and the upper continental slope. These accumulations are usually unstable, and can move downslope as slumps, slides, and debris flows, leaving scars on the seafloor and the marks of downslope transportation and deposition of sediments. Factors known to influence these catastrophical events include loading or oversteepening of slopes, underconsolidation of sediments due to rapid sedimentation, earthquakes, gas build-up and sea-level changes (Saxov and Nieuwenhuis, 1982). Mass-movement deposits are often called debris, and correspond to mixed lithologies with hard pebbles and boulders, with sizes up to several tens of metres. They will show up in sidescan sonar images as bright reflective areas (see Figure 7.6) with fine-grained textures.

GLORIA imagery from the Western Atlantic continental rise presents several good examples of debris slides. In particular, Figure 7.9 presents the junction between two slides of different acoustic characters. This image results from the mosaicking of several swaths, and ranges in depths between -4,250 m and -5,000 m. Cape Fear slide is recognised by its high backscatters and the very fine-grained textures all along the slide. Individual flows are difficult to distinguish because most lack well-defined edges (at least at GLORIA's resolution). Only the two larger ones (upper-left portion of the image) are visible, with medium-bright backscatters. The Cape Fear slide involved slope failure of an area 37 km long, 10 to 12 km wide, and up to 80 m thick (EEZ-Scan, 1991). Debris flows from the scar area can be traced across the continental rise for over 250 km as a bright reflection return. At the NE edge of this image, debris from the Cape Fear slide flow across earlier debris from the Cape Lookout slide, indicating that the Cape Fear slide is more recent. It is also marked by a bright acoustic return across the rise. Within the bright return area, the bottom exhibits a fuzzier texture than the Cape Fear slide, and sometimes looks as more "transparent". This is also characteristic of mass-movement debris as imaged by GLORIA: 3.5-kHz profiles show the "transparent" aspect is related to the deposition of finer sediments on top of an acoustically harder layer.

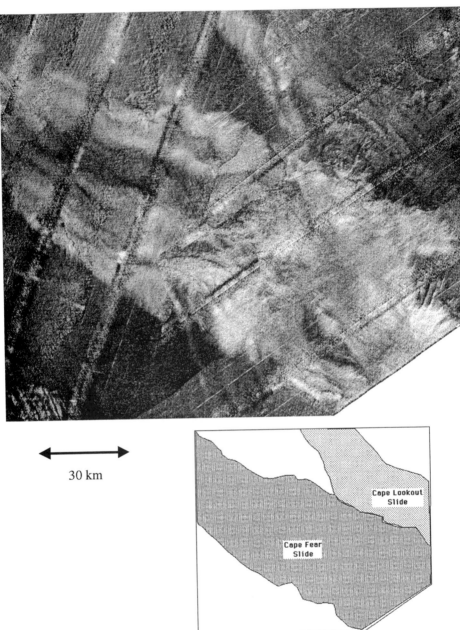

30 km

Figure 7.9. Cape Fear and Cape Lookout slides. These slides are about 60 km and 40 km wide, respectively. GLORIA imagery, courtesy United States Geological Survey.

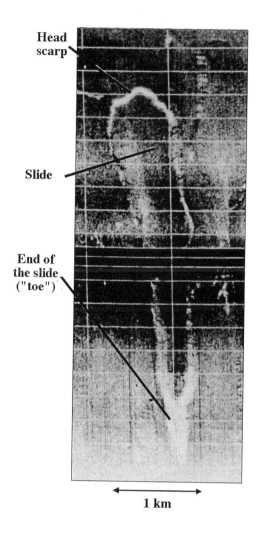

Figure 7.10. Teardrop slide imaged by SeaMARC I. The image has been inverted so that high backscatters are bright, low backscatters are dark. From O'Leary and Dobson, 1992.

Mass-movement features include debris tongues or rubble trails that extend from the bases of the larger scarps, and small slides located along the surface of the lower slope ramps. Some of them have a teardrop form similar to land-based features (called "disintegrating soil slips" in California). The image presented in Figure 7.10 has been acquired with SeaMARC I near the Block Canyon slide, on the Northeastern USA continental margin (O'Leary and Dobson, 1992). SeaMARC I was a 30-kHz system with a resolution of 2.5 metres and a swath width of 5 km. This deep-tow sonar was the ancestor of SeaMARC II (see Table 2.1) but never saw much use. The sonar track is visible as a horizontal black line in the middle of the image. The arcuate head scarp is visible in the upper portion, with a bright reflectivity because it is facing the sonar. Its relief is of the order of 10 metres. The outlines of the slide are brighter than their surrounding, their grey levels apparently not changing with distance to the sonar track. This can be explained by a higher roughness of the seafloor at the contact between the young slide and its substratum. The slide is 800 m wide and 4,100 m long, going downslope. The toe forms a thin, tapered tail of rubble, with a uniform high backscatter. The presence of teardrop slides in the slopes is often used as an evidence of recent slope unstability.

Other types of turbiditic deposits are associated with canyon overflows, when the incoming fluxes of sediments are too large or too rapid to pass through the canyon. Overflows are more likely to happen near meandering canyons. Levees and overbank deposits form low sediment ridges along the banks of the canyons (Figure 7.11). They can be a few hundreds of metres wide, and up to several metres high. They can also form in several episodes, and create terraces (Figure 7.12). These terraces generally extend perpendicular to or even downslope from the canyon.

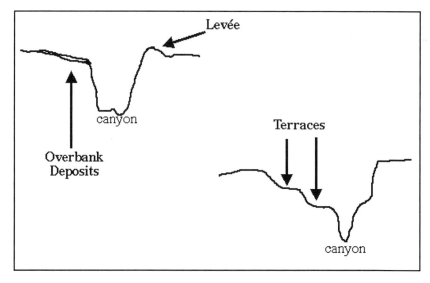

Figure 7.11. Stylised topographic profiles across a canyon, showing different types of turbiditic deposits.

Figure 7.12 presents SeaMARC II imagery from the continental margin off Peru and Chile. The original data (upper image) has been inverted in order to represent the highest backscatter values as white and the lowest backscatter values as black (lower image). The ensonification direction is constant through the image. The homogeneous area in the middle of the image corresponds to a data gap, when the sonar stopped recording on both channels for a short time. The terraces appear as contiguous areas of grainy texture, located on the sides of the canyon axis.

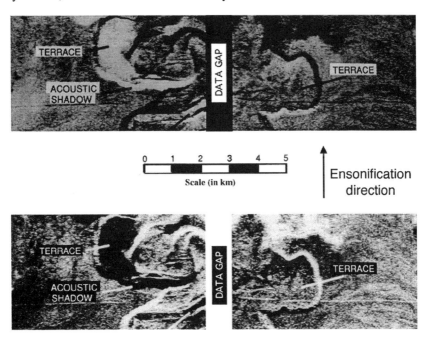

Figure 7.12. Terraces imaged by SeaMARC II in the Southern Pacific. The lower image is the same than the upper image, but inverted so that grey levels are directly proportional to backscatter values. Courtesy R. Hagen, AWI (Germany) and Elsevier Publications.

7.2.4 Sediment Redistribution

Bottom currents rework and redistribute the deposited sediments and form large, elongated structures in the areas with little or no topographic variations. These structures are sediment waves, similar to subaerial sand dunes. They will be recognisable in sidescan sonar imagery from the contrasting acoustic returns between the slopes facing toward and away from the sonar. Depending on the local conditions (speed, direction and longevity of bottom currents, surrounding topography), sediment waves will have different sizes and shapes. SeaMARC II imagery from the Peruvian continental margin shows arcuate to linear sediment waves between the canyons (Figure 7.13). These waves vary in orientation and have wavelengths of 400 to 800 metres. Acoustic shadows are small or absent, hinting that the topographic expression of sediment waves is very small. In this case, echo-sounder profiles show it to be less than 5 metres. Individual wave crests can be traced for up to 6 km on the images. In this region, the sediment waves are grouped into distinct wave fields with common orientations and wavelengths.

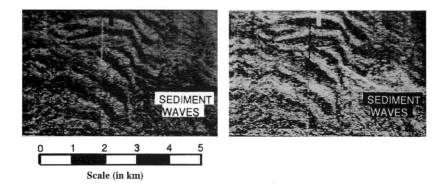

Figure 7.13. Arcuate sediment waves, ensonified from the bottom of the image. SeaMARC II imagery is generally represented with grey levels inversely proportional to the backscattering amplitude (left). An inverted image is shown at right, with high backscatters bright and low backscatters dark. Courtesy R. Hagen (AWI, Germany) and Elsevier Publications.

Sediment waves exist at all scales. The smaller ones are no more than metre-scale ripples, similar to the ones seen in the next chapter (Chapter 8: Coastal Environments). The larger ones can extend for several tens of kilometres (Figure 7.14). This image was recorded with GLORIA in water depths of 4,750 m, in the middle of a flat basin with no large variations in topography. The large sediment waves, known as the Lower Rise Hills, are clearly seen from the contrasts between the slopes facing away from and toward the sonar. Closely spaced, they are several tens of kilometres long and up to 5 km wide. These sediment waves are located on the flank of the Hatteras Ridge, and are believed to have been formed by the interaction of contour-following currents, the northeast-flowing Gulf Stream, and the southwest-flowing abyssal Western Boundary Undercurrent.

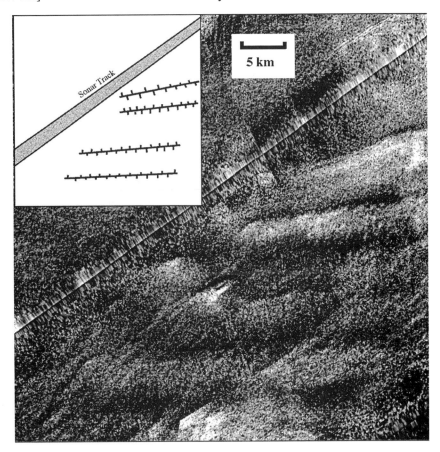

Figure 7.14. Linear sediment waves, imaged by GLORIA in the continental rise off North Carolina, Eastern USA. Courtesy United States Geological Survey.

Sediments deposited along channels can be redistributed through overflow or slumping. The TOBI record presented in Figure 7.15 is a raw image from the NW African continental rise, and corresponds to a half-swath (approximately 3 km). The image has not been slant-range corrected, which prevents accurate measurements of across-track distances. Three domains can be drawn diagonally. The upper left one corresponds to the sediment channel: backscatter levels are bright, and the fine-grained texture follows the direction of the channel. Acoustic shadows thrown by the channel sides are well marked. Measurements of their across-axis lengths yields estimates of 50 to 60 metres for the height. This is confirmed by deep-tow 7.5-kHz profiles across the channel. The lower portion of the image is the largest one in area, and corresponds to flat sediment deposits. Backscatters are bright to medium bright and the textures homogeneous. They are separated from the steep-sided channel by elongated "ribbons" forming structures analogous to small sand waves. The largest of these "ribbons" are approximately 1 km long and 30 m wide. Their acoustic shadows indicate heights of the order of 20 m above the surrounding seafloor. The "ribbons" decrease in size as they are closer to the channel. These structures are interpreted as slumps, i.e. the results from downslope movement of loose sediments. One possible cause of submarine slumps is the undercutting of channel walls by turbidity currents. Other causes include over-loading or earthquakes.

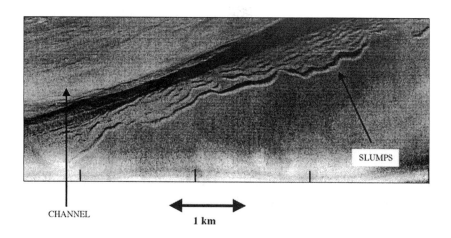

Figure 7.15. Uncorrected TOBI imagery, NW African continental rise. The vehicle track is along the bottom of the image. The channel is 1-2 km wide, 50 m deep, and can be traced for over 100 km on GLORIA mosaics from the whole region. Courtesy D. Masson, SOC (UK.)

7.3 TECTONIC STRUCTURES

On continental margins, the geological structures which can be directly attributed to tectonism are very rarely seen on the surface. They are generally highly sedimented, and visible only on subsurface profiler data or on seismic profiles. Nonetheless, some fresh tectonic expressions can be imaged with sidescan sonar. They will appear exactly as their analogues at mid-ocean ridges (see Chapter 5). The slopes facing toward the sonar will be reflective to highly reflective, whereas the slopes facing away from the sonar will be less reflective or completely dark.

The image presented in Figure 7.16 was acquired with the Russian sonar MAK-1 in the Mediterranean Sea, and inverted to show higher backscatter levels as brighter tones. The sonar track is horizontal and in the middle of the image. A single long structure, about 300 m wide, cuts the image diagonally from the lower-left to the upper-right corner. This structure shows consistent backscatters, very bright in the upper half of the image, medium dark to medium bright in the lower half, where it is associated with small gullies close to the sonar track. All this argues for an interpretation as a fault scarp, sloping down from the top-left to the bottom-right. The small circular structures visible in the rightmost part of the image are thought to be pockmarks, i.e surface manifestations of gas seepage (see section 7.7.3). Slight along-track variations of backscatter are visible, and are attributed to the lack of appropriate angle-varying gain corrections.

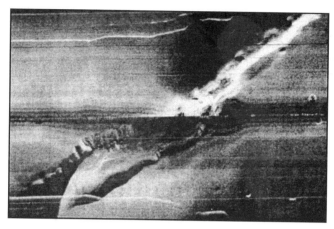

Figure 7.16. Uncorrected MAK-1 record of a fault scarp in the Mediterranean Sea. The image corresponds to a full MAK-1 swath and is 2 km wide. From Beijdorff et al., 1994. Copyright UNESCO.

7.4 EXAMPLES OF REGIONAL IMAGERY

7.4.1 The Blake Escarpment, North Atlantic

Continental margins have been more extensively surveyed than other regions on the seafloor, for example abyssal plains (see Chapter 6). It is difficult to select examples of regional imagery among all the data available, and we have restricted ourselves to only two types of frequency, and two types of geological environments. The first example presented was acquired with GLORIA during the survey of the Eastern US Exclusive Economic Zone (EEZ SCAN, 1991). Several GLORIA swaths were mosaicked to produce an image of the Blake Escarpment, a spectacular submarine cliff marking the transition between the continental margin and the abyssal plains (Figure 7.17(a)). In less than 15 km, the transition is particularly abrupt between the intermediate depths of the Blake Plateau (-2,500 m) and the abyssal depths of the Blake Basin (-4,900 m). The top of the Escarpment is visible throughout as a bright acoustic reflector, but its base is only visible as the boundary with the homogeneous sediment patches of the Blake Basin.

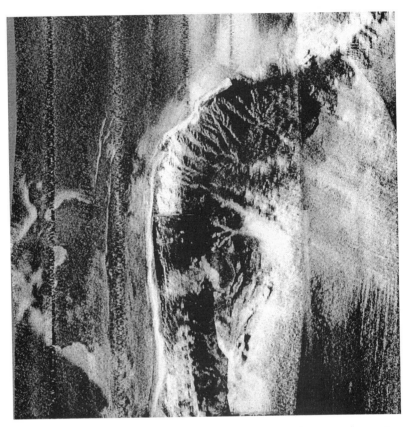

Figure 7.17(a). GLORIA regional imagery of the Blake Escarpment, on the continental margin east of Florida. The image is 51 km wide and 51 km long, oriented with North on top. Depths vary abruptly between 2,500 m in the left half of the image and 4,900 m in the right half. Courtesy United States Geological Survey.

The Blake Plateau is just west of the Blake Escarpment (Figure 7.17(b)), and is made of sediments with a homogeneous medium-bright backscatter interpreted as large-scale sediment drift deposits. Long fine lineations are visible close to the Escarpment, parallel to it and extending for several tens of kilometres. They are probably caused by erosion and minor slumping of the deposits. The dark featureless region in the SW of the image is very similar to regions of Figure 7.3, and is confirmed by coring as being a patch of carbonate mud.

In the extreme north-east of the image, a very bright zone extends north-eastward and is very narrow. This is the steepest part of the escarpment. Only the stronger echoes generated near the top of the escarpment are returned. Submersible dives show that the cliff is near-vertical over much of its extent. The strong bottom currents following the lower contours scoured the seafloor. They created moats, visible as dark linear structures, at the junction with the Blake Basin.

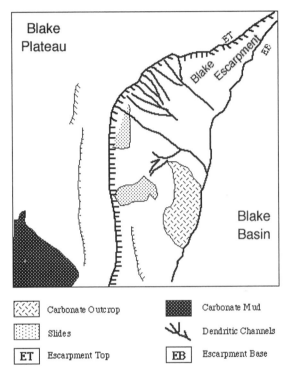

Figure 7.17(b). Interpretation of the GLORIA image of Figure 7.17(a).

The main portion of the image, in the centre, shows that the Blake Escarpment is extensively cut by short dendritic channels merging into canyons. The bands of medium-bright backscatter on top of the Escarpment are other indicators of sediment drift deposits, like the ones further west in the Blake Plateau. They are present all along the escarpment, but are only a few kilometres wide at most. Their narrowness is used in some interpretations to argue for a smaller sediment supply than necessary to form the canyons, and for their formation in very shallow waters (in the Early Cretaceous, 100 Ma BP). The larger portions of the Blake Escarpment are also marked by sediment slides, with a more mottled acoustic texture (Figure 7.17(b)). The southernmost of these slides seems to stop abruptly along a large structure, around which canyons also seem to overlap. This suggest that this structure is fairly old, at least older than the canyons and slides, and resistant to erosion by the incoming turbity flows. Seafloor photographs and samples led to the conclusion this was a large carbonate outcrop.

7.4.2 The East Arequipa Basin, South Pacific

Submarine canyons formed on passive continental margins (such as the Eastern USA) are thought to form by mass-wasting and erosion on the continental slope. They show distinctive patterns similar to fluvial drainage patterns. Submarine canyon systems at active margins are more complex, as they are subject to tectonic activity (uplift or subsidence) and higher rates of sedimentation. This second example of regional imagery was recently acquired on the continental slope off Peru with the high-resolution sidescan sonar SeaMARC II (Hagen et al., 1994). This active continental margin is close to the Peru-Chile trench, formed by subduction of the Nazca Plate beneath the South American Plate (see Chapter 4: Deep-Ocean Trenches and Collision Margins). Its broader portion is 170 km wide, and lies at depths between 400 m and 1,600 m. It contains the heavily sedimented Arequipa Basin, from about 16°S to 18°30'S.

Sidescan sonar imagery of the Eastern Arequipa Basin reveals the existence of a large submarine canyon system with several tributaries (Hagen et al., 1994). The main branch of the canyon extends over 160 km from the shelf break off southern Peru to its termination in the trench off northern Chile. The highly meandering canyon is 350 m to 1,100 m wide, and has relief of 150 to 250 m through most of its distance. Figure 7.18 shows the central portion of the East Arequipa Basin submarine canyon. SeaMARC II imagery is traditionally represented with grey levels inversely proportional to backscatters, which explains the unusal appearance of some of the structures. The image is formed by the mosaic of two slightly overlapping E-W swaths. Their junction, in the middle of the image, is marked by abrupt changes in backscatter levels, corresponding to the juxtaposed northward and southward ensonifications of topographic structures. The canyon is the main feature visible in this image. Intricately meandering, it has an overall sinuosity of 1.95, comparable to highly meandering subaerial rivers. Meandering is apparently controlled by the basin slope and the characteristics of the turbidity currents that form the canyons. Cut-off meanders are visible as thinner loops on the edges of the canyon axis. They bound terraces spaced regularly, at several levels above the canyon floor. Sediment waves are visible on the east of the canyon. Already shown in section 7.7.4, these waves have wavelengths of 400 to 800 metres and very low topographic amplitudes. The wave fields are more pronounced toward the convex downslope meanders and are absent on the western side of the canyon. This is interpreted as evidence of their formation by turbidity currents overflowing from the canyon (Hagen et al., 1994).

Figure 7.19 follows the submarine canyon south, as it is joined by smaller tributaries. This image is made up of three E-W swaths. The abrupt changes in backscatter levels associated to the stitching are particularly visible on the canyon axis near 18°17'N, and with the sedimentary deposits in the NE corner of the image. The canyon is more sinuous, and only one cutoff meander is visible, larger than its northern analogues. Three separate, narrower canyons merge with the main canyon at 18°15'S. Its width increases with the larger volume of sediments to transport. The depth varies between 1,200 m at the top of the image and 2,000 m at the bottom. Again, sediment waves are located on the eastern sides of the canyons. The western sides are marked by a patchy sediment deposition with alternating streaks of higher and lower backscatters oriented downslope. A small slump (< 1 km^2) in the NE corner of the image is associated with the channel of one of the smaller canyons. Its setting is similar to the slump seen in Figure 7.16, and the origins may be the same.

Figure 7.18. Sidescan sonar imagery of the central portion of the East Arequipa Basin submarine canyon. SeaMARC II imagery shows high backscatters with dark tones, and low backscatters with bright tones. Courtesy R. Hagen, AWI (Germany) and Elsevier Publications.

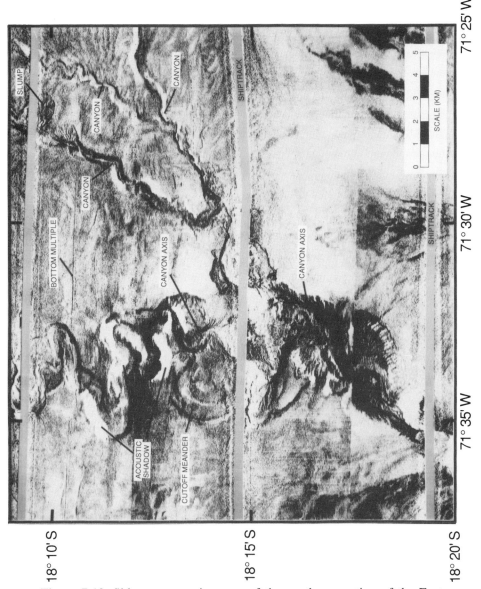

Figure 7.19. Sidescan sonar imagery of the southern portion of the East Arequipa Basin submarine canyon. SeaMARC II imagery shows high backscatters with dark tones, and low backscatters with bright tones. Courtesy R. Hagen, AWI (Germany) and Elsevier Publications.

7.5 VOLCANIC STRUCTURES

Continental margins are generally far from areas of volcanic productivity, such as mid-oceanic ridges or subduction zones. Very few volcanic structures have been documented in continental margins, and most if not all of them are inactive and highly sedimented. When still intact, these volcanoes look very much like their counterparts at spreading centres (see Chapter 5: Mid-Ocean Ridge Environments). Their dimensions will be more important, to emerge above the sediment layers. For example, Bear Seamount in the New England seamount chain (Figure 7.20) is very similar to the W-seamount on the Mid-Atlantic Ridge (Figure 5.7).

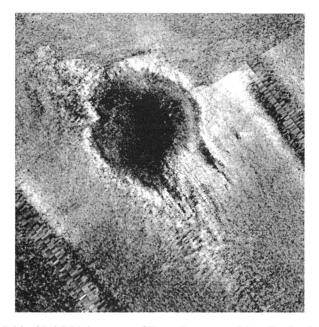

Figure 7.20. GLORIA imagery of Bear Seamount, New England seamount chain. The image is oriented North. The volcano is nearly circular, 14 km wide, and 1,500 m high. Courtesy United States Geological Survey.

Located off Georges Bank on the Eastern USA continental margin, this slightly sedimented volcano appars as a large circular structure with a diametre of 14 km. The stitching of the two survey lines has been carefully performed, and the seam is barely visible on the median line between the swaths. The noisy areas below the nadir are slightly widening toward the bottom-right corner of the image. This means that the distance between the sonar instrument and the ground is increasing. Because GLORIA is shallow-towed, it can therefore be assumed that the topography is dipping moderately toward the bottom-right. This "educated" guess is confirmed by concurrent bathymetric maps of the area, which show that depths vary between 2,750 m in the NW corner of the image and 3,250 m in the SE corner.

The large, circular shape of the seamount is divided into a bright (reflective) half and a dark half, corresponding to the direction of ensonification. The extent of the shadow on the left (several kilometres) indicates that the height of the structure is of the order of a few hundred metres. The reflective area on the right can be separated into a flat semi-circular rim, and a brighter, more homogeneous slope facing directly toward the sonar. Bright textural patterns, associated with debris from the volcano, can be seen all around. Most seamounts have been affected by tectonics and erosion processes. Knauss Knoll (Figure 7.21) is another inactive seamount at the edge of the Eastern USA continental margin. GLORIA imagery reveals an asymmetric shape elongated along-track, with dimensions of 6 km by 10.5 km. The slopes of the seamount close to the sonar track are highly reflective (nearly saturated). A small circular shape on top of the slopes is associated with a small shadow, and is interpreted as the summit crater. The seamount's extremity further from the sonar track is less reflective. It delimits an embayed slope, indicative of an extensive collapse of the original edifice. Adjacent seafloor sediments overlap the seamount flanks and attenuate the backscatter in the regions which are not directly facing the sonar. They scoured the seafloor and created a small depression (dark backscatter patch SW of the volcano). Analysis of dredge samples gives an age estimate of 100 Ma to 85 Ma. Knaus Knoll is typical of the other seamounts in the New England chain, and, to a certain extent, is also typical of relict seamounts in continental margins.

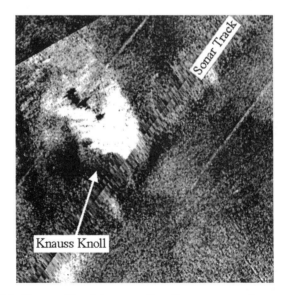

Figure 7.21. Knauss Knoll is a small sediment-covered seamount on the Eastern USA continental margin. Its dimensions are approximately 6 km (along-track) by 10.5 km (across-track). GLORIA imagery, courtesy United States Geological Survey.

7.6 BIOLOGICAL ACTIVITY

With the right combination of frequency and ground resolution, sidescan sonar imagery can reveal some manifestations of biological activity. On continental shelves, slumps of large algae and local accumulations of dead and living plants are the most likely objects to be detected. For example, direct visual observations of *Halimeda* plants on the Indonesian Margin show that living stands of these plants organise into groups with common reliefs of 0.5 m above the seafloor. In shallower waters, coral structures are also visible (see Chapter 8: Coastal Environments). These structures are quite different from their sedimentary background and will generally present stronger backscatters.

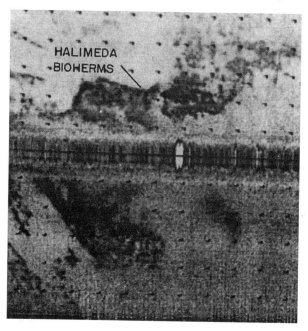

Figure 7.22. Sidescan imagery showing *Halimeda* bioherms on a flat sedimented bottom. The black crosses are spaced evenly every 25 metres. The image comes from a paper record; high backscatters are dark, low backscatters bright. Unpublished data, courtesy M. Veerayya and V. Purnachandra Rao, National Institute of Oceanography, India.

Figure 7.22 shows the accumulation of *Halimeda* in a mound (bioherm), on a flat portion of the western Indian continental shelf. This image was acquired with the SMS-960 sidescan sonar (frequency 110 kHz) in water depths close to 90 metres. Backscatters are inverted, and are represented by dark levels when higher. The main part of the seafloor exhibits a medium backscatter, light grey on one side of the swath to dark grey on the other side (due to different calibration settings). This substratum corresponds to a portion of a drowned carbonate platform covered by sediments. The highly reflective patches (dark in Figure 7.22) have unusual shapes. They are formed of several distinct units, round and a few metres wide. These units are interpreted as individual *Halimeda* bioherms, composed of clustered dead and living plants, which is confirmed by coring. Echo-sounder profiles confirm that these units formed by coalescence of several mounds, 2 to 14 m high.

Acquired during the same survey, a larger-scale image shows the different morphologies of *Halimeda* bioherms (Figure 7.23). Morphologies range from numerous small mounds (a few metres in diameter) through "haystack" elongated features (several tens of metres wide, several hundreds long) to broad swells in the upper-right corner of the image. Bioherms on the shelf-edge appear more linear (Figure 7.24), whereas bioherms inside the platform generally assume mound-like morphologies. Both types form by biological accumulation, and these shapes are related to the dynamics of *Halimeda* growth and clustering. The lineated aspect of the algal ridges in Figure 7.24 suggests an orientation along the main direction of the bottom currents which bring the necessary nutrients.

Sidescan sonar provides information about biological activity which supplements the results from echo-sounder or seismic profiles, coring, and *in situ* observations. The large-scale mapping of biological activity is only possible with sidescan sonar, and repeated surveys can assist in the long-time environmental monitoring of these regions. It can also supply information about palaeo-climates. In this case, for example, the *Halimeda* bioherms are shown by coring to date from about 10,000 years BP, and to have originated in an arid climate (Rao et al., 1994). Although rare, and very dependent on the combination of frequency and resolution chosen for the sonar, biology-oriented sidescan surveys form an important asset in these studies.

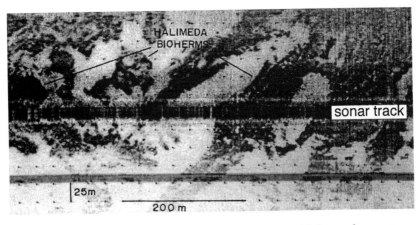

Figure 7.23. Sidescan imagery of massive algal biohermal structures, separated by sediment-covered areas of the Indian continental shelf. High backscatters are dark, low backscatters bright. Courtesy V. Purnachandra Rao and M. Veerayya, National Institute of Oceanography, India, and Elsevier Publications.

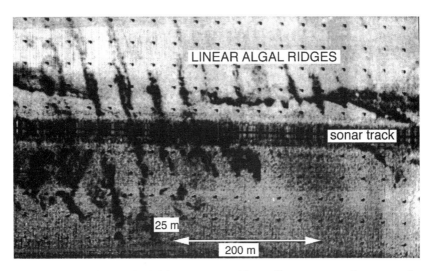

Figure 7.24. Linear algal ridges separated by sediment-covered areas on the carbonate platform off Bombay, western continental shelf of India (M. Veerayya, pers. comm., 1995). The water depth is 90 metres. High backscatters are dark, low backscatters bright. Unpublished data, courtesy V. Purnachandra Rao and M. Veerayya, National Institute of Oceanography, India.

7.7 STRUCTURES FROM THE EPICONTINENTAL SEAS

7.7.1 Mud Volcanism

Mud volcanism has been documented in various deep-sea environments all over the world. The most-studied examples are located in the Mediterranean and in the Black Sea. Other occurrences have been documented in accretionary complexes like the Barbados, where mud volcanoes have been actively formed over at least the last 200,000 years, the Nankai accretionary prism and the Indonesian arc (see Chapter 4). Mud volcanism may also appear in coastal environments, like the Makran region in Pakistan, where active mud volcanoes are found in association with gas seeps on- and off-shore. Their mechanisms of formation are diverse. In the Black Sea, mud volcanoes are formed by the channellised eruption of gas and fluid-saturated clays through thick overlaying sediments (Limonov et al., 1996). Conversely, mud diapirs in the Mediterranean are the result from lateral tectonic compression pushing upward plastic material.

Mud volcanoes appear on sidescan sonar imagery as highly reflective patches of seafloor on a dark background. They generally have a mushroom or cone-like shape, with diametres at their base of a few hundred metres to 3 km, and rise from 20 to 150 metres above the seafloor. The bright patches present irregular boundaries and a backscatter generally much stronger than its surroundings. This stronger backscatter is considered to be caused by the irregular relief of the crater and cone, and by the presence of mud breccia and rock blasts on the seafloor or at very shallow depths. It seems that mud diapirs are not visible on GLORIA imagery if the pelagic sediment cover is greater than about 2 metres (Fusi and Kenyon, 1996): the highly reflective circular patches can only be interpreted as areas of substantial micro-scale roughness. This indicates that the frequency and the ground resolution (respectively 6.5 kHz and 50 m for GLORIA) play an important role in the detection or non-detection of the structures associated with mud volcanism.

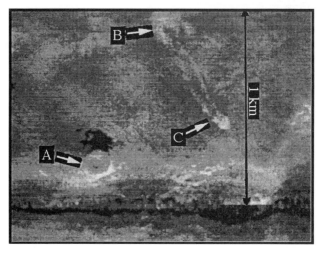

Figure 7.25. Mud vents and mud flows in the Mediterranean, imaged by MAK-1: (A) Jaén Dome; (B) possible mud vent; (C) single, small mud flow. From Ivanov et al., 1996. Copyright Elsevier Publications.

Jaén Dome, in the Mediterranean, is an example of a very small mud volcano (Figure 7.25). This image presents a half-swath of a survey with the Russian sonar MAK-1 (see Table 2.2). Jaén Dome itself appears as a small circular structure, approximately 150 m large. It corresponds to a small high on the seafloor, as shown by the reflective half-rim facing the sonar and the cone-shaped shadow. There is a small crater at its centre, and there are no traces of mud flows along its flanks. The main backscatter values are higher than the surroundings, but not strongly. Bottom sampling of this structure showed the presence of mud breccia (creating the higher intensity because of its roughness), below about 2 m of hemipelagic sediments (attenuating the sidescan sonar signal). This type of mud volcanoes is thought to be characterised by either weak eruptions, or by very viscous eruptive products which cannot form distinctive flows.

Two other structures are discernible in Figure 7.25. The first one is a possible mud vent in the far range (noted "B"), characterised by an irregular elliptical patch with a higher reflectivity. Originating from it is an elongated feature, with brighter boundaries and a circular bright end, apparently higher than its surroundings (presence of a brighter rim facing the sonar). This structure is very reminiscent of small debris slides (see section 7.2.3), but it outlines a smooth texture. It is interpreted as a single, small mud flow, extending for several hundred metres. Generic studies have shown that flow structures and mud ridges can appear independently of the domes, although always in their vicinity.

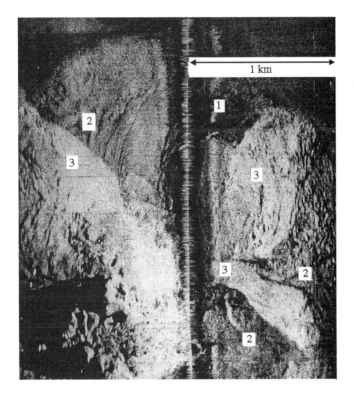

Figure 7.26. MAK-1 imagery of the Yuzhmorgeologiya mud volcano in the Black Sea. Mud flows are referenced from the oldest (1) to the youngest (3). They can be distinguished by their respective textures and backscatter levels. From Ivanov et al., 1996. Copyright Elsevier Publications.

Yuzhmorgeologiya is one of the largest mud volcanoes found in the Black Sea (Figure 7.26). The whole structure has a diameter on the seafloor of more than 2.5 km, and a relative height of 100 to 150 metres with gentle slopes. The MAK-1 imagery shows a full swath, slant-range corrected. Several cones are visible on the highly backscattering dome in the lower-left corner of the image, projecting important shadows. Their dimensions range from a few tens down to a few metres. At least three generations of mud flows are visible, originating from this dome. They can distinguished by their respective textures and backscatter levels. The youngest flows (denoted "3") are highly reflective (nearly as much as the dome itself), with mottled textures. The older flows (denoted "2") exhibit medium backscatters, with a smoothed texture where heterogeneities are still visible. And the oldest flows (denoted "1) are nearly dark and completely homogeneous. Ground-truthing relates these differences in backscatter and texture to variations in the thickness of the overlaying pelagic sediments (i.e. age), and changes in the physical characteristics of the mud breccia. The relative importance and extensions of the overlapping mud flows show that the Yuzhmorgeologiya mud volcano developed probably over a long time, and that its activity gradually decreased.

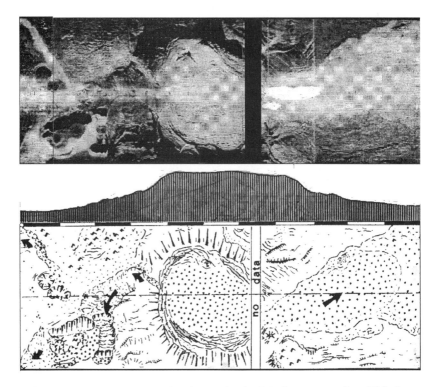

Figure 7.27. Bergamo mud volcano, in the Mediterranean Sea. This image was acquired by the EG&G990S sidescan sonar (its track is horizontal in the middle of the image). The portion presented here is 800 m large and 2,200 m long (the scale division for the interpretation is of 200 metres). The volcano's shape (in middle) is derived from acoustic profiler data. Image and interpretations by Hieke et al., 1996. Copyright Elsevier Publications.

Bergamo Dome (Figure 7.27) is another mud volcano from the Mediterranean Sea. It is a small dome, less than 1 km in diametre. It was imaged with the EG&G990S sidescan sonar (see Table 2.2) from an altitude of 100 m above the seafloor. The track line is horizontal and crosses the volcano in its centre. The mud volcano is characterised by a very strong backscattering, outlining the two circular rims on each side of the sonar and their shadows. This wrinkled bulge is interrupted on the right by a highly reflective lobe, extending downslope and becoming larger. This lobe displays in its interior and its margin faint lobate structures, suggesting a debris flow which is creeping over an old lobe coming from a neighbouring mud volcano. It is bounded on its northern and southern sides by higher walls. On the northwestern slope of Bergamo Dome, dark bands of high reflectivity also suggest downflowing mud breccia. The flow apparently starts from a source at the edge of the central plateau (indicated by an arrow on the interpretation). This flow branches in two arms flowing around some circular structures of a disk-shaped relief, which are all approximately 100 m in diameter. Two or three of these structures show in their centres small depressions, about 50 m in diameter. These depressions may well be other small mud cones. These different interpretations are summed up in an interpretive 3-D sketch of Bergamo Dome (Hieke et al., 1996), showing the relation of the different structures with one another (Figure 7.28).

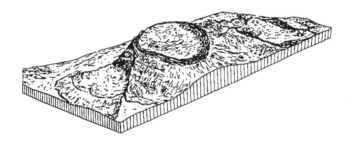

Figure 7.28. 3-D interpretation of Bergamo Dome. From Hieke et al., 1996. Copyright Elsevier Publications.

In conclusion, mud volcanoes and mud cones look remarkably similar in their broader characteristics: they present high reflectivities, very different from their surroundings, and have circular to elliptical shapes.The largest volcanoes also present radial mud flows. Ground-truthing shows that the variations in backscatter intensity across different mud volcanoes and mud flows are caused by variations in thickness of the pelagic deposits overlaying the mud breccias. Changes in the physical properties of the mud breccia (density, sizes and importances of clasts) also cause variations in reflectivity, as the result of both volume and surface roughness backscatter. The amount of gas saturation varies from one region to the other (it is for example higher in breccias from the Black Sea mud volcanoes) and contributes to the total backscatter signal, if the frequency of the sonar is close to the resonant frequency of the inclusions.

Mud volcanism and the associated structures are very important scientifically, because they provide unique information about the geological processes taking place at larger depths through the study of vented gases and muds. They are also interesting as potential traps for hydrocarbons, and because they create important modifications of the seafloor, which are potential hazards for telecommunication cables and pipelines.

7.7.2 Brine Accumulation Structures

Structures related to local brine accumulation can be found in various environments, and manifest themselves as bottom depressions filled with hypersaline water. Brine pools have been particularly well documented in the Mediterranean, but their exact modes of formation are still poorly understood. They are thought to result from the rapid dissolution of evaporites (salt deposited by the evaporation of seawater, in this case during the drying of portions of the Mediterranean in the Late Miocene).

The differences in physical characteristics (salinity, density, temperature) of the hypersaline water and normal sea water create severe acoustic contrasts between the brine pools and their surroundings. The incoming sound waves are reflected and refracted on the different layers that normally constitute the pools. The reflected waves interacted with a very smooth interface, and therefore none of their energy is scattered back to the sonar sensor. In the absence of backscattered sound, the pools' surfaces appear uniformly black. Around the edges of the pools, the brine layers are thinner, and constructive and destructive interference from the refracted waves give the edges a locally bright appearance.

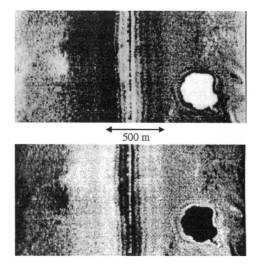

Figure 7.29. Brine pool imaged by MAK-1. Top: original image. Bottom: the same image, inverted to present grey levels directly proportional to backscatter. From Woodside and Volgin, 1996. Copyright Elsevier Publications.

Some dark, echo-free patches were discovered during a survey of the Eastern Mediterranean mud volcanoes (Figure 7.29), in water depths of 2,050 m and more. The deep-tow sonar, MAK-1, was flying at 100 m above the seafloor. The anechoic patch does not present any variation with the distance from the sonar track, i.e. the incidence angle. The grey levels are consistently at 1 or 2, compared to grey levels of 20-30 for the surrounding hemipelagic sediments, and grey levels of 100-120 for the neighbouring mud volcanoes between which the survey line was run. It is therefore most unlikely that this patch be an outcrop of sediments with different physical properties, as they would still produce some backscattering. The shape is sub-circular, with a diametre of about 260 metres. A few rings of slightly different backscatter intensities are visible around it, presumably where the brine layers are thinner.

1 km

Figure 7.30. L'Atalante Basin, imaged with TOBI in the Mediterranean Sea. The large black structure corresponds to the brine lake. The higher values in the left arms of the basin are artefacts introduced during processing. Courtesy G.K. Westbrook, University of Birmingham (UK.)

L'Atalante Basin is a large brine lake, with dimensions of several kilometres (Figure 7.30). It was imaged with TOBI during several passes, which have been merged together in this image. The lake itself is in the lower part of the image, and appears uniformly black. Some higher values in the left arms were produced by the amplification of ambient noise by TVG during processing (see Chapter 2: Sonar Data Processing). They correspond to the overlapping of the far ranges from both the left swath and the right swath. The edges of the lake, further away from the sonar track, are highly reflective. This is explained by the interference of acoustic waves when the brine layers become thinner. The brighter linear portion in the middle of the basin may be interpreted as a fault scarp facing the sonar. The TOBI image and the multibeam bathymetry concurrently available show that some of the margins of the lake are linear and semiparallel, which indicates that the form of the basin is partly controlled by faults. North of the brine lake, four small depressions in the seafloor are present, with sizes ranging from 100 to 1,000 metres, with a round to irregular form. They have an appearance similar to pockmarks caused by the escape of gas (see section 7.7.3). Temperatures and chemical anomalies were sampled inside L'Atalante Basin. They showed the brine lake is organised in three layers, with respective thicknesses of 16 m, 30 m, and 40 m. The temperature in each layer increases gently downward, and is higher than in the surrounding seawater. L'Atalante Basin's brine is also very rich in potassium.

Brine lakes and pools are uncommon features on the seafloor, and have been discovered in the proximity of mud volcanoes or of important faults. They have sometimes been associated with tectonically induced widespread fluid circulation through evaporites. Researchers now use them to constrain models for the formation and evolution of mud diapirism, as well as for the study of past and present salinity budgets in the deep ocean. Brine pools also present a biological interest, since the recent discovery of deep-sea communities apparently based on chemosynthesis.

7.7.3 Pockmarks and Seepages

Pockmarks are sedimentary features associated with past or present seepage of gases and fluids. They were first seen on sidescan sonar records in the 1970s, and can be found in the epicontinental seas as well as on continental margins around the world (Hovland and Judd, 1988). Pockmarks are formed in three stages: (1) building up of gas or pore-water pressure in a porous layer of sediments below an impermeable cohesive (excess pressure is relieved by doming of the seabed); (2) after fracturing of the dome, excess gas pressure is released by eruption, along with the overlying sediments; (3) the fine-grained sediments become suspended in the water and transported elsewhere by currents, while coarser material falls back into or near the newly formed pockmark.

Most pockmarks are approximately circular in plan, but there is a considerable variety in shapes and sizes. Because they are active features, their dimensions are likely to vary over time. They correspond to hollow depressions of the seabed, a few metres deep. Their appearance on sidescan sonar imagery will therefore be sub-circular or elliptical, the slope facing away from the sonar being darker and the slope facing the sonar brighter. This will of course vary according to the angle of incidence and how far the pockmarks are from the sonar.

A sidescan sonar survey of the Eratosthenes Seamount (Eastern Mediterranean) shows some circular structures interpreted as pockmarks (Figure 7.31). They are visible on both sides of the sonar track (horizontal line in the middle of the image). These features are characterised by a diameter of some tens to 300 metres, and their relative depth is several metres. These holes are notable by high reflective rims on the side facing the MAK-1 sonar. Many of them have a quite perfect circular shape. The ones north of the sonar track are more elliptical and exhibit a stronger backwall reflection, apparently due to the local slope of the seamount (which goes up northward). The holes could be related to karst, but more likely they originated from shallow gas and/or fluid seeps and represent pockmarks.

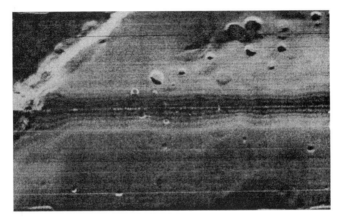

Figure 7.31. Pockmarks on top of Eratosthenes Seamount (Eastern Mediterranean), imaged with MAK-1. The image is 3 x 2 km. From Beijdorff et al., 1994. Copyright UNESCO.

A comprehensive study of pockmarks, their dimensions and their settings, was performed by Hovland and Judd (1988). They distinguish several generic varieties:

- *standard pockmarks and elliptical pockmarks:* these are the most common. They are shown in Figure 7.31, and some of them were visible around the L'Atalante Basin (Figure 7.30).

- *composite pockmarks:* they occur where individual pockmarks merge with one another. In some instances, groups of pockmarks are found clustered together, while in other cases the merging has proceeded to the extent where a single feature with a complex shape is all that is left.

- *asymmetric pockmarks:* on sidescan sonar imagery, they show a distinct and quite often long tail, and a strong backwall reflection on one side only. This is due to the local slope up to seabed level, long and gentle.

- *pockmark strings:* they extend for several hundred metres, and are made up of small pockmarks, symmetrical, shallow, and 10 to 15 metres in diameter. Pockmark strings observed in the Norwegian Trench (North Sea) were attributed to pre-existing iceberg ploughmarks which focused the gas seepages.

- *elongated pockmarks and troughs:* they are so elongated that they resemble gullies or troughs. The upper sediment layers are absent, and the older sediments are exposed. An example of elongated pockmark is given in Chapter 8.

- *unit pockmarks:* these seabed depressions are very small (< 5 m) and shallow. They are also referred to as "pits" or "pit clusters", and seem to be incipient pockmarks.

Because of their origin (fluid or gas seepage) and of their locations (very often close to hydrocarbon reservoirs), pockmarks generate a considerable interest in off-shore industries, in particular those related to gas exploration and exploitation. This interest is enhanced by the consequences pockmarks may have on off-shore platforms and the security of drilling exploration. Catastrophic gas escapes can form large, deep pockmarks in very short periods of time. In one occurrence, cited by Hovland and Judd (1988), a crater 600 m across and 30 m deep was formed within a period of 5 days. In another example, the gas blowout at a depth of 240 m reached to the surface, and instabilities led to the abandonment of a neighbouring drilling platform. Even if these examples are quite extreme, it nonetheless demonstrates the possible consequences of drilling into shallow gas pockets, or what could happen to the stability of platforms in the vicinity of such an event.

7.8 SUMMARY

The images presented in this chapter aim at presenting the different processes encountered in continental margins. They are focused on the transport of sediments from the coast to the abyssal plains, through channels and slides, and the effects of this transport on the seafloor (e.g. deposits, erosion). Images from wide-swath sonars such as GLORIA are privileged because of the important size of most structures present in the continental slopes. Smaller-scale structures (mainly sedimentary) are very similar to the ones observed in other environments, and only the most unusual or typical ones were described thoroughly. Tectonic and volcanic activity is rare, and only a few occurrences are shown. Examples of regional imagery were drawn from a passive continental margin, in the Eastern USA, and from an active continental margin disappearing into a subduction zone, off Peru and Chile.

Certain structures are typical from continental margins or from epicontinental seas such as the Mediterranean. Rarely published manifestations of biological activity are presented in the chapter, along with some features recently discovered: mud volcanoes, brine pools. A particular importance is attached to the acoustic recognition of pockmarks and seepages, because of their obvious economic significance. All these examples were acquired using sidescan sonars as varied in frequency and resolution as GLORIA, SeaMARC II, TOBI, EG&G990S and MAK-1.

Emerged lands are a large provider of sediments to the continental margins, and the coastal environments are presented in the next chapter. Although marginal in size, coastal waters are very important in terms of economy, strategy and environment. They also provide the largest opportunities for sidescan sonar surveys. As the water depths decrease to a few tens of metres at most, the deployment techniques change slightly, and so does the interpretation. Geological or not, the structures visible near the shores are generally closely related to their subaerial equivalents. Two friends and colleagues, internationally known for their contribution to the field of shallow-water sonar imagery, have agreed to present this next chapter.

7.9 FURTHER READING

- **About continental margins in general:**

Poag, C.W., P.C. de Graciansky; *Geologic Evolution of Atlantic Continental Rises*, Van Nostrand Reinhold, New York, 378 pp., 1992

EEZ-SCAN 87 Scientific Staff; "Atlas of the U.S. Exclusive Economic Zone: Atlantic continental margin", United States Geological Survey, Miscellaneous Investigations Series, vol. I-2054, 174 pp., 1991

- **About submarine slides:**

O'Leary, D.W., M.R. Dobson; "Southeastern New England continental rise: Origin and history of slide complexes", in *Geologic Evolution of Atlantic Continental Rises*, C.W. Poag and P.C. de Graciansky (eds), Van Nostrand Reinhold, New York, 378 pp., 1992

Saxov, S., J.K. Nieuwenhuis; "Marine slides and other mass movements", Plenum Press, New York, 1982

- **About brine pools:**

Westbrook, G.K., and the MEDRIFF Consortium; "Three brine lakes discovered in the seafloor of the Eastern Mediterranean", EOS Trans. AGU, vol. 76, n. 33, p. 313-318, 1995

- **About pockmarks and seepages:**

Hovland, M., A.G. Judd; "Seabed pockmarks and seepages: impact on geology, biology and the marine environment", Graham & Trotman, London, 293 pp., 1988

- **About the sonar observation of biological activity:**

Roberts, H.H., C.V. Phipps, L.L. Effendi; "Morphology of large *Halimeda* bioherms, Eastern Java Sea (Indonesia): a side-scan sonar study", Geo-Marine Letters, vol. 7, no. 1, p. 7-14, 1987

Rao, V.P. M. Veerayya, R.R. Nair, P.A. Dupeuble, M. Lamboy; "Late Quaternary *Halimeda* bioherms and aragonitic faecal pellet-dominated sediments on the carbonate platform of the western continental shelf of India", Marine Geology, vol. 121, p. 293-315, 1994

- **About mud volcanism in the Mediterranean and Black Seas:**

Marine Geology Special Issue, June 1996

Limonov, A.F., J.M. Woodside, M.K. Ivanov (eds); "Mud volcanism in the Mediterranean and Black Seas and shallow structure of the Eratosthenes Seamount", UNESCO Reports in Marine Science no. 64, 173 pp., UNESCO, 1994

- **About sediment transport processes:**

Hagen, R.A., D.D. Bergersen, R. Moberly, W.T. Coulbourn; "Morphology of a large meandering submarine canyon system on the Peru-Chile forearc", Marine Geology, vol. 119, p. 7-38, 1994

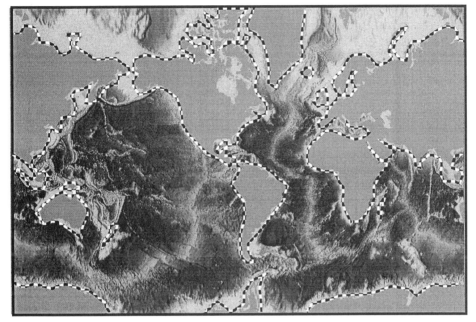

Figure 8.1. Coastal environments throughout the world.

8

Coastal Environments

Invited contribution from Doris Milkert and Veit Hühnerbach
University of Kiel, Germany

Coastal environments are essential to our everyday life. They shape and are shaped by our commercial, ecological, and cultural activities. Most of the world's fishing is drawn from coastal waters, particularly in developing countries. Coastal management and construction of dykes, platforms, coastal defenses, etc., influence near-shore processes and *vice versa*. Sediment redistribution in estuaries has historically led to the abandonment or modification of once prospering harbours (e.g. Brugge in Belgium). Industrial activities, especially the more environmentally detrimental ones, have marked effects on the seafloor and its associated biology. Together, these activities indicate the importance of coastal environments to society. The successful maintenance of coastal areas, their waterways and harbours, requires a precise knowledge of sediment circulation patterns. This information is readily available from sidescan sonar imagery.

8.1 GEOLOGICAL BACKGROUND

The exact geographical extent of coastal environments is difficult to define (Reading, 1978). Generally, they comprise areas of the seafloor that are shallower than 200 metres, reasonably close to the shore, but with a wide range of hydrographic settings. They are liable to rapid changes, influence and are influenced by, lacustrine, riverine and estuarine environments. For the purpose of this book, coastal environments are loosely defined here as the transition zone between land and sea, down to a depth of 200 m (Figure 8.1). For completeness, lakes and inshore water bodies have also been included in this chapter.

Chapter 7 (Continental margins) discussed the effects of the enormous input of sediments from coastal areas into the deeper ocean. In shallow seas, the seafloor is mostly covered by sediments, exhibiting a great variety of types and bedforms. Unconsolidated sediments usually cover more solid substrates, although, in some circumstances, relict basement structures remain. Active processes, such as currents, waves and glaciations, erode basement structures and reshape existing sediment patterns (e.g. ripples and iceberg ploughmarks). Biological activity is widespread in all coastal environments, often creating prominent structures such as coral mounds, algal mats, seagrass fields and macrophytes. These features, which can all be observed on sidescan sonar, reveal the increasing anthropogenic influence on this fragile environment. Most of the examples presented in this chapter are drawn from western European waters. However, they are typical of, and equally applicable to, coastal environments throughout the world.

8.2 SEDIMENTARY COVER

In shallow shelf seas, the hydrodynamic and geological settings generally control the pattern of sediment distribution. The decrease in wave energy with greater water depth is often reflected in coastal sediments by a general decrease in sand-grain size and an increase in mud content (Newton and Stefanon, 1975). This has a major effect on sediment distribution in many coastal areas, often resulting in a heterogeneous seafloor.

In high-latitudes, coarser sediments in shallow water are often associated with material from the last Quaternary glaciation. Foe example, moraines and till can be found in many coastal areas around the North Sea and the Baltic. Figure 8.2, an unprocessed sonar record, was acquired at a frequency of 100 kHz in a water depth of 18 m. It shows a substrate, with moderate acoustic backscatter, upon which are superposed sub-circular objects with distinctive shadows. The objects range in size from a few centimetres to several metres. The length of their shadows, combined with a knowledge of the altitude and range of the sonar, reveals object with heights ranging from of 1.8 m (e.g. at the top of Figure 8.2) to ~30 cm. The distinctive, rounded, shape and size of the objects identify them as pebbles, cobbles and boulders, lying on a substrate of glacial till. This image is of a relict Quaternary moraine, in which the coarser sediments have been left on the surface by wave and current erosion. Additional information from the area shows that this type of relict sediment occurs in patches or stripes, 10 - 30 cm thick, and is often covered with algae and sessile animals. In this example, the till is thinly covered with coarse lag sediments, alternating with coarse or medium grained sand. This thin cover forms a heterogeneous region of backscatter, with an organised and grainy texture, corresponding to ripple marks and finer-grained sediments.

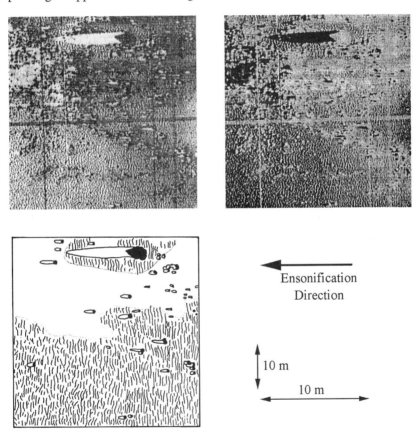

Figure 8.2. The 100-kHz sonar image (Klein Model 590) shows a typical example of a coarse lag sediment that thinly covers eroded glacial till. The original image (from the western Baltic Sea in a water depth of 18 m) is shown on the upper-left side. The upper-right image is an inverted view in which low backscatter is dark (the convention adopted for this book). An interpretation is given in the adjacent line drawing. Courtesy of the Geological-Paleontological Institute, University of Kiel, Germany.

The shallower parts of the coastal environment are largely influenced by the action of waves and strong currents. Residual sediments of gravel are frequently found at shallow depths in near-coastal areas. They are particularly abundant off-shore from retreating cliffs or on submarine hills, but are frequently interspersed with patchy sand areas. This patchiness implies that sediments of different grain sizes (pebbles, gravel, coarse sand and medium to fine sand) are simultaneously exposed to wave and current action. The contact between coarse sediments and small rocks reflects the local changes in the slope and hydrographic energy.

Figure 8.3 was recorded at a frequency of 500 kHz in a water depth of 15 m and an altitude of 5 m above the seafloor. The image covers an area of 40 x 75 m. It presents a mixed area: one with a heterogeneous texture of high backscatter, and another of low, homogeneous backscatter. The area of low backscatter corresponds to mud, and appears similar to mud imaged with GLORIA on the continental shelf (Figure 7.3). High backscatter is ascribed to coarser-grained sediments.

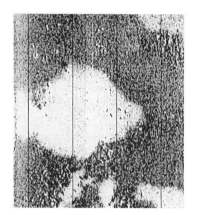 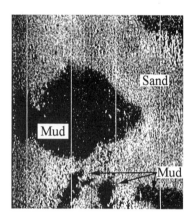

Figure 8.3. 500-kHz sidescan sonar record (Klein Model 590) showing different sediment types and patchy structures. The original image is shown on the left hand side, whereas the image on the right hand side is inverted. Ensonification is from left to right and the black vertical lines are spaced every 10m. Courtesy of the Geological-Paleontological Institute, University of Kiel, Germany.

Undisturbed, organic-rich, muddy sediments, with a large water content, appear to high-frequency sonar as almost featureless areas with low backscatter. Figure 8.4 is a 500-kHz sonar image, recorded in the central parts of a mud-filled channel in the Baltic Sea. The sonar fish was towed 8 m above the seafloor in a water depth of 25 m. The image covers an area of approximately 50 square metres. In this example, very low backscatter is caused by shallow gas within the uppermost sediment. Good images from muddy sediments are difficult to obtain because of acoustic energy absorption and reflection away from the sonar. The best images are obtained with high to very high acoustic frequencies. The only distinct features that can be distinguished on Figure 8.4 are two otter-board tracks in the center of the image.

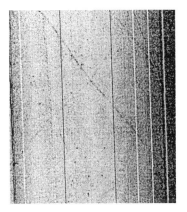 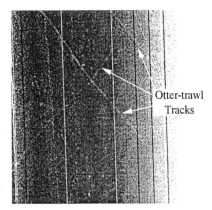

Otter-trawl
Tracks

Figure 8.4. This 500-kHz sonar image (Klein Model 590) shows almost featureless muddy sediment from the inner parts of a glacial shaped fjord in the western Baltic Sea. In the center of the image, otter-board tracks are visible. Black vertical lines are spaced every 10m. The original image is given on the right-hand side and the inverted image on the left. Courtesy of the Geological-Paleontological Institute, University of Kiel, Germany.

8.3 ROCK OUTCROPS

The shape and distribution of outcropping rocks and rocky bottoms are easily detected with sidescan sonar. Figure 8.5 shows outcropping Devonian "Old Red" sandstone on the seafloor of Liefdefjord, Spitsbergen (northern Norway). This image was recorded at 100 kHz, in a water depth of ~20 m, at an altitude of 10 m above the seafloor. The scene can be divided into several oblique bands. Very regular patterns are visible with alternating reflective stripes and accompanying acoustic shadows. These stripes have a width of up to 8 m and diagonally span the entire field of view. With their high backscatter and relatively smooth textures, these features are an excellent example of stratified bedrock. From nearby land exposures, it is known that the strata reach a thickness of 2-3 m, and have a dip of ~70°. The lower part of the image is occupied by smoother, less reflective textures, that are interpreted as sand. The along-track variations in backscatter result from manual changes of the gain by the sonar operator. Because the image is unprocessed, the progressive widening of the nadir region towards the bottom of the image shows a small but steady increase in vehicle altitude.

Because of the differences in their formation and structure, outcropping metamorphic rocks often appear different to the other types of outcrop. The unprocessed, 500-kHz, sonar image presented in Figure 8.6 shows an example from Roscoff, NW France. This image has been ensonified from the top, with the sonar fish towed at 15 m above a flat seafloor. The entire water column is visible at the top of the image as a region of no backscatter. With its homogeneous texture and a medium backscatter, the surrounding seafloor is interpreted as sandy. Large objects, visible in the right-hand part of the image, include one that protrudes into the water column. This object, therefore, is much shallower than its surroundings. The first returns on the image represent the profile of the seafloor, from which it can be seen that the outcrop stands proud by 7 m. The great variability in backscatter is due to scour marks and furrows on the surface of the rocks and the surrounding seafloor. By comparing them with nearby features on-shore, these large objects are interpreted as outcropping metamorphic rocks. Shadows correspond to local, small-scale, variations in topography.

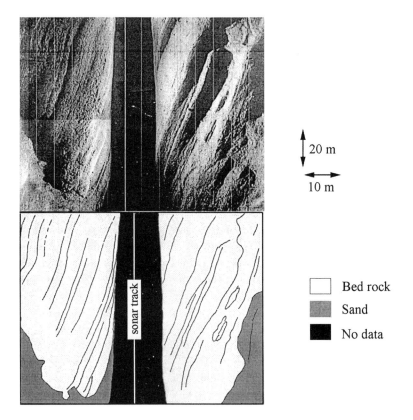

20 m

10 m

☐ Bed rock

▨ Sand

■ No data

Figure 8.5. Unprocessed, 100-kHz sidescan sonar record (Klein Model 590) of Devonian, stratified, bedrock outcropping in the Liefdefjord area, Spitsbergen. The strata strike NNW-SSE and are depicted on the interpretation as thin lines. White vertical lines are spaced 10 m apart. Courtesy of R. Schacht, Geological-Paleontological Institute, University of Kiel, Germany.

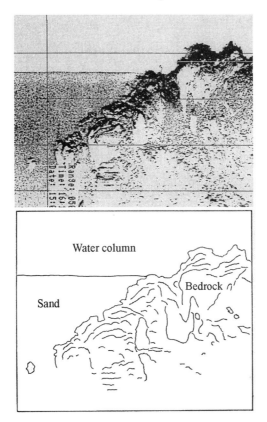

Figure 8.6. Analogue 500-kHz sidescan sonar image (Klein Model 590) of outcropping metamorphic bedrock near Roscoff, NW France. Black horizontal lines are spaced 10 m apart. Strong backscatter is dark. Courtesy of W. Tietze, University of Marburg, Germany and A. Wehrmann, Forschungsinstitut Senckenberg, Wilhelmshaven, Germany.

During burial, sediments are often consolidated and compacted. Later erosion can re-expose them. The sedimentary nature and degree of compaction of these features affects their appearance on sidescan sonar. Figure 8.7 is a 100-kHz sidescan sonar image, acquired in water depths of 15 m, close to the German island of Sylt. Parallel variations in backscatter on the left side of the image, combined with alternating, sharp crests and shadows, are interpreted as current-induced ripples. These ripples are interrupted by a sharp edge that delimits a feature with lower backscatter. Ground-truthing of this low-backscatter block has revealed it to be an outcrop of consolidated, Eemian clay, that can be traced to neighbouring land exposures.

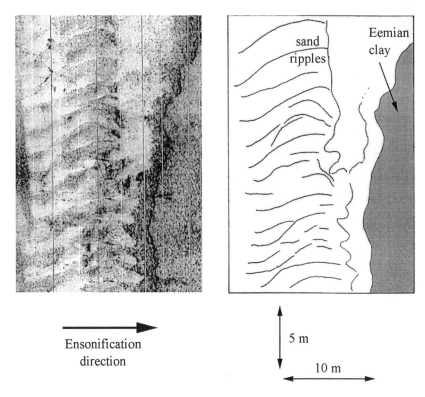

Ensonification
direction

5 m

10 m

Figure 8.7. 100-kHz sidescan sonar image (Klein Model 590) of outcropping Eemian clay and sandy sediments off Sylt Island (Germany). Strong backscatter is dark. The sonar was towed 5 m above the bottom. Current-induced ripples can be seen located close to the edge of the more consolidated clay. Black vertical lines are spaced 10 m apart. Courtesy of K. Schwarzer, Geological-Paleontological Institute, University of Kiel, Germany.

8.4 SEDIMENTARY FEATURES CREATED BY WAVES AND CURRENTS

Sediments are routinely and extensively reworked and redistributed, especially in shallow water environments, by high tidal and wave energies. Some coastal areas experience semi-diurnal tides with a range of 3 to 4 m, and maximum surface current speeds ranging from 0.6 to 1 m s^{-1}. As a result, tidal shelves can exhibit a wide range of bedforms that are in dynamic equilibrium with their hydraulic environment. Distinctive bedforms and sedimentary facies, such as straight crested sandwaves, furrows, gravel waves and sand ribbons, are characteristic of transport by tidal currents.

8.4.1 Ripples

Ripples are the simplest bedforms associated with sediment redistribution. They occur on non-cohesive surfaces (i.e. sands) as undulations, oriented transverse to the direction of water flow. Ripples can be occasionally found on muddy seafloors. Produced by the interaction of waves or currents, ripples are conventionally described in terms of their size and shape. Their generic classification includes: wave ripples, current ripples (transverse), isolated ripples (incomplete) and combined current/wave ripples (see Reineck and Singh, 1980 for a review).

Typical wave-induced ripples, from the shallow parts of the western Baltic Sea, can be seen in Figure 8.8. During this 100-kHz survey, the sonar was towed 6 m above the seafloor, in a water depth of 18 m, close to the island of Helgoland in the North Sea. The image can be divided into 2 major elements: dark homogeneous patches with low backscatter, that are interpreted as mud (and are similar to those in Figure 8.3); and linear, parallel backscatter variations that correspond to small wave ripples with both bifurcated and regular crests.

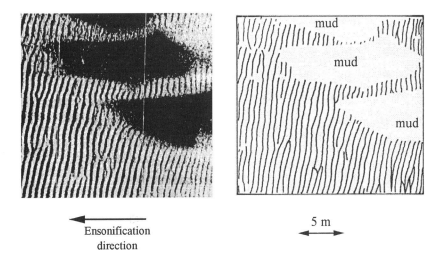

Ensonification
direction

5 m

Figure 8.8. 100-kHz sidescan sonar image (Klein Model 590) showing wave ripples in medium-grained sand showing both bifurcated and regular crests with intervening patches of mud. Low backscatter is dark. Courtesy of F. Werner, Geological-Paleontological Institute, University of Kiel, Germany.

Depending on grain size and current intensity, the width of the individual ripples can vary from a few centimetres to several metres. The length of the ripples is constrained by the influence of currents, waves and the local topography. The example shown in Figure 8.9 was acquired at 500 kHz, in 18 m of water. Several fields of ripples can be seen on a background of sand with a medium backscatter. The left-hand field is about 10 m wide.

The right-hand field is 20 to 30 m wide and divided into several regions of ripples with various crest spacing. Their formation in the bottoms of channels restricts the lengths of the ripples. Small, 10 m scale, oscillations of the nadir (bottom part of the image) are not related to topographic variations, rather to movement of the sonar vehicle as a result of a light wind and swell during data acquisition.

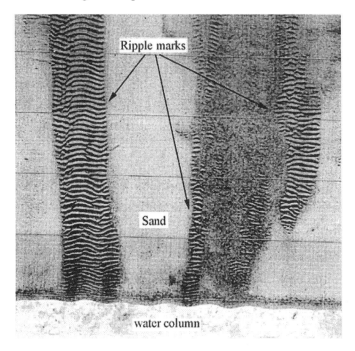

Figure 8.9. 500-kHz sidescan sonar image showing channels filled with sandy sediment that has been sculpted into a set of ripples. Horizontal black lines are spaced every 10 m. In this analogue, paper record, backscatter is proportional to the darkness of the image. Courtesy of the Geological-Paleontological Institute, University of Kiel, Germany.

8.4.2 Megaripples

Megaripples are found in areas of relatively high-current velocities (0.6 to 1.2 m s^{-1}) and water depths greater than 18 m. They vary in height from 3 to 9 m, and from 125 to 1300 m in length. Megaripples are often covered with transient, smaller ripples of the type described above. Figure 8.10 shows the juxtaposition of megaripples (also called 'giant ripples') with small ripples on their flanks. This unprocessed image was collected at a frequency of 100 kHz in a sandy tidal-flat close to Sylt Island (Germany). It shows broad variations in backscatter, characteristic of megaripples. Darker lineations between the megaripples may indicate a new ripple generation. Regular variation of the depth to the first return is correlated to the topography of the megaripples. Close examination of the center of the image shows secondary ripples on the flank of one of the megaripples. Ground-truthing has shown that the seafloor is uniformly composed of terrigenous sand.

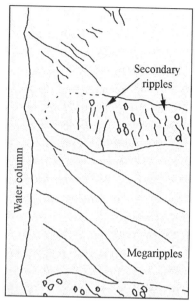

Figure 8.10. Uncorrected 100-kHz sidescan sonar image (Klein Model 590) of sandy megaripples in a tidal-influenced area off northern Germany. Ensonification is from the left and strong backscatter is bright. Vertical white lines are spaced 10 m apart. Courtesy of the Geological-Paleontological Institute, University of Kiel, Germany.

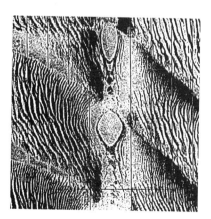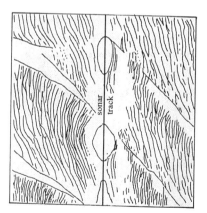

Figure 8.11. Dowty Widescan 3050 sidescan sonar image of megaripples composed of fine carbonate sand (Roscoff, France). The sonar was towed in a water depth of 25 m. Black vertical lines are spaced 10 m apart. Courtesy of B. Bader and P. Schäfer, Geological-Paleontological Institute, University of Kiel, Germany.

The composition of the ripples does not influence their overall morphology. Figure 8.11 shows megaripples formed from carbonate sand (in this case, particles of bryozoan shells). Located in a tidally influenced area off western France, this 325-kHz sonar image shows two megaripples. Large current ripples with a crest spacing of 0.8 to 1.4 m are visible between the megaripples, with smaller ripples on their flanks. The high contrast of the image is a result of manual gain adjustments, made to accommodate the large range in backscatter. Uncorrected slant-range results in "bull's-eyes" in the nadir over topographic lows between the megaripples.

8.4.3 Longitudinal Bedforms

Finer-grained sediments, suspended and redeposited by bottom currents, may erode coarse and consolidated substrates. This is similar to the action of turbidites in deep water (Figure 6.15). Figure 8.12, recorded with a 100 kHz sonar, shows current-induced ribbons of fine- to medium-grained sands. They are surrounded by much coarser lag sediments comprising eroded till, very coarse sand and gravel. Lineated variations in backscatter indicate longitudinal furrows, caused by erosion from suspended sediments, and longitudinal ridges of unconsolidated sand. The strike of the lineations indicates the direction of bottom currents.

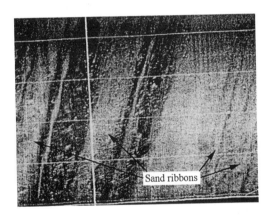

Figure 8.12. Analogue, 100-kHz sidescan sonar image (EG&G Mk. 1), from the Baltic Sea. The image is ensonified from the bottom to the top and strong echoes are dark. Longitudinal sand ribbons of fine-medium sand are formed by bottom currents. They overlie much coarser lag sediments that comprise eroded till, very coarse sand and gravel. White horizontal lines are spaced 10 metres apart. Courtesy of the Geological-Paleontological Institute, University of Kiel, Germany.

8.4.4 Obstacle Marks

Characteristic sedimentary structures form around large objects subjected to bottom currents. These structures, classified as crescents, shadows and moats, can be seen in both deep and shallow water (see Chapter 6: Abyssal Plains and Basins, and Chapter 7: Continental Margins). Their recognition and analysis reveals information about local current conditions. Such data are crucial in areas such as the North Sea where it is necessary to install drilling rigs and oil-production platforms on current-swept sandy areas.

Figure 8.13 shows a 500-kHz image of scours and moats in the Baltic Sea. The sonar was towed 5 m above the seafloor in a water depth of 15 m. The triangular region of high backscatter, in the center of the image, is a channel containing mainly coarse sand. Isolated circular objects, with distinct haloes, are interpreted as cobbles lying on surrounding finer-grained sediment. Haloes of lower backscatter around the cobbles indicate asymmetric moats which have been scoured by storm-induced bottom currents. The sidescan image, which was acquired at a towing speed of 5 knots, has not been anamorphically corrected, resulting in an apparent along-track stretching of the objects.

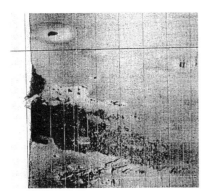 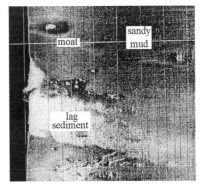

Figure 8.13. Unprocessed, analogue image (Klein Model 590) of current-induced scours and moats around cobbles on a seafloor of sand and mud (Baltic Sea). Ensonification is from the left. The original record is shown on the left with strong echoes as dark. The inverted image is shown on the right. Along-track stretching of the image is a result of poor anamorphic correction for the sonar's speed. Black vertical lines are spaced 10 m apart. Courtesy of K. Schwarzer, Geological-Paleontological Institute, University of Kiel, Germany.

8.5 GLACIAL FEATURES

8.5.1 Iceberg Ploughmarks

Almost nine tenths of an iceberg lies below sealevel, and these "keels" have a variety of shapes and sizes. When they ground, these various keel shapes leave characteristic marks on the seafloor. Erosional features, related to the influence of glaciers, are a well-known phenomenon from most ice-influenced environments around the world. Various names have been applied to such features including: ploughmarks, scores, ice-gauges, grooves and furrows (Elverhoi, 1984, Hambrey, 1994).

The most common marks are multiple, parallel scour channels bordered by levees. These can reach widths of 20 to 100 m and depths of 2 to 10 m. Not confined to coastal environments, they have been observed on seafloors as deep as 950 m (e.g. the flanks of the Iceland-Faroe Ridge). Figure 8.14 was acquired in a water depth of 100 m near Spitsbergen. The sonar was towed 18 to 20 m above the seafloor. This 500-kHz image reveals lineated features of high and low backscatter. They are interpreted as scours and pits formed by wallowing icebergs. The sudden change in backscatter towards the bottom of the image was caused by a manual change of the recording gain.

 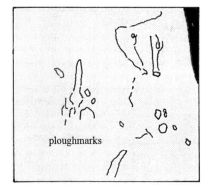

Figure 8.14. Unprocessed 500-kHz sidescan sonar image (Klein Model 590) showing iceberg plough-marks in a glacial fjord (Spitsbergen, Norway). The area shown on the image is about 40 x 50 m. Ensonification is from the right and strong echoes are dark. White lines crossing the image are print-head artefacts. Courtesy of R. Schacht, Geological-Paleontological Institute, University of Kiel, Germany.

8.5.2 Slump Structures in the Arctic Environment

In the glaciomarine environment, over-steepened slopes of poorly consolidated sediments are often unstable. This results in the widespread occurrence of submarine slides, slumps and gravity flows (Hambrey, 1994). These features have a hummocky, sometimes lobate, morphology (see Chapter 7: Continental Margins). In coastal areas, they frequently have a relief of several metres and extend laterally over hundreds of metres. Figure 8.15 shows a 500-kHz sidescan sonar record of slump structures from an Arctic glacial fjord (Spitsbergen, Norway). Changes in backscatter are largely a result of the relief of the slump. The sonar was towed 20 m above the seafloor in a water depth of 80 m. The example is taken from a location close to the front of an active glacier. Black lines crossing the image are caused by noise from a 3.5-kHz seismic survey that was run simultaneously with the sidescan sonar.

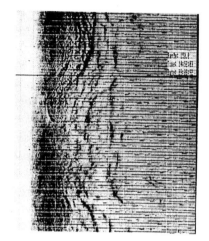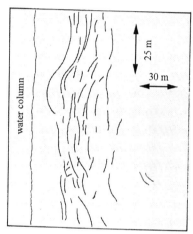

Figure 8.15. Unprocessed Klein Model 590 image (500-kHz) showing slump structures in the Arctic environment of a glacial fjord system, northern Spitsbergen. Ensonification is from the left and strong echoes are dark. Black lines crossing the image are caused by noise from a 3.5-kHz seismic survey. Courtesy of R. Schacht, Geological-Paleontological Institute, University of Kiel, Germany.

8.6 POCKMARKS

Pockmarks are sedimentary features caused by the escape of gas or fluids through the seafloor (Hovland & Judd, 1988). They have been imaged by sidescan sonar in many environments around the world, including continental margins (see Chapter 7). High-resolution sidescan sonar has been used for studying the distribution, geometry and internal structure of pockmarks in the Eckernförde Bay, Baltic Sea (Figure 8.16). The water depths in the bay reach 26 m. During acquisition of the image, the sonar was towed 7 m above the seafloor. Linear features in the center of the image are otter-board tracks (see section 8.8.2). The unprocessed 500-kHz image shows a slightly lighter and irregular-shaded patch, roughly 180 m in length and 40 m wide. Its lower backscatter, compared with the surrounding mud, is interpreted as a result of high gas content in the sediment. This 1 to 2 m deep feature has a coffer-like cross-section, elongated shape and an irregular outline. It is interpreted as a pockmark. Although the origin of the Eckernförde Bay pockmarks is still debated (e.g. Kandriche and Werner, 1995), it is probable that many result from the release of hydrocarbon gas from the seafloor. However, pockmarks in the outer parts of the bay may be associated with the release of fresh water from an artesian aquifer.

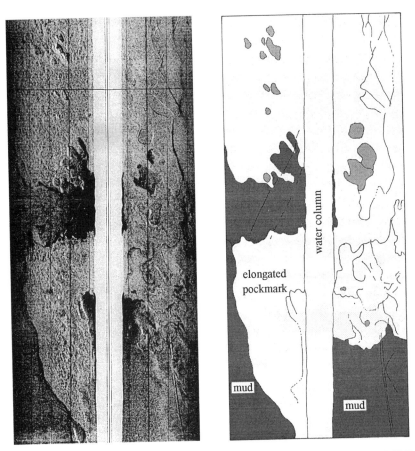

Figure 8.16. Unprocessed 500-kHz sidescan sonar image (Klein Model 590) showing an elongated pockmark from Eckernförde Bay, Baltic Sea. Black vertical lines are spaced 10 m apart. Strong echoes are dark. An interpretation, shown on the right-hand side indicates, the major features including the linear otter-board tracks. Courtesy of F. Werner, Geological-Paleontological Institute, University of Kiel, Germany.

Figure 8.17 is a composite of several sidescan swaths, all acquired at 100 kHz in a water depth of 80 m. During the survey a number of parallel tracks were recorded from 15 m above the seafloor and stencilled manually. The dotted lines on the interpretation denote boundaries between sidescan swaths, requiring care when interpreting images like this. Figure 8.17 shows typical, circular pockmarks (diameter 10 to 20 m) on the floor of lake Bodensee, Germany. The image also shows a funnel-shaped pockmark (centre left), caused by gas escaping through the upper sediments. Small features at the centre of the pockmarks are interpreted as craters caused by sediments blown clear of the seafloor by gas flow (see Hovland & Judd, 1988). In this area, limnic, calcareous mud surrounding the pockmarks produces a homogenous backscatter.

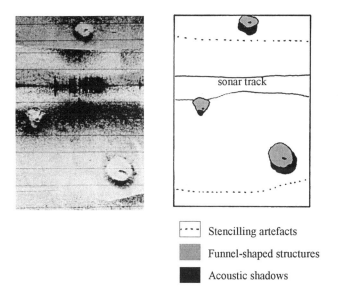

Stencilling artefacts

Funnel-shaped structures

Acoustic shadows

Figure 8.17. Unprocessed, analogue, 100-kHz sidescan sonar record (Klein Model 590) of pockmarks on the floor of Lake Bodensee (Germany). Black horizontal lines are spaced 10 m apart. Strong echoes are dark. Courtesy of H.G. Schröder, Limnological Institute, Langenargen, Germany.

Small funnel-shaped structures on the floor of another small lake in northern Germany are shown in Figure 8.18. The sonar was towed at 12 m above the seafloor, in 22 m of water. The features appear as low-backscatter circles (ranging in diameter from 3 to 4 m) with dark spots in their centers. Like the ones shown in Figure 8.17, these circular features are interpreted as pockmarks with central craters. Although they are smaller, more abundant and often aligned, their origin is probably similar to those in Figure 8.17. Small ripple-shaped darker features surround the pockmarks, and are interpreted as fine- and coarse-grained sediments, swept by currents.

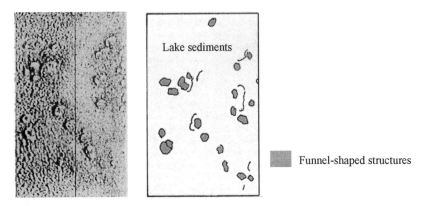

Figure 8.18. Unprocessed, analogue, 100-kHz sidescan sonar record (Klein Model 590) showing funnel-shaped structures on the floor of a small lake (Germany). These are pockmarks with an average diameter of 3 to 4 m. Courtesy of the Geological-Paleontological Institute, University of Kiel, Germany.

8.7 BIOLOGICAL ACTIVITY

8.7.1 Corals

Although corals are typical of shallow water environments and warm climates, they also occur in deep water (between 200 m and 300 m) at higher latitudes. These deep-water corals have been recognized for a long time (see Henrich et al., 1995, for a recent review). Recently, it has become increasingly important to gain a better knowledge of these ecosystems. Rich in biomass and calcium carbonate, they provide a large reef-forming potential in the aphotic zone.

Sidescan sonar imagery was used to survey deep-water reefs on the Sula Ridge (Norway). Figures 8.19 and 8.20 are images acquired with a 500-kHz system that was towed at 15 to 20 m above the seafloor in a water depth of 280 m. Both images show important circular structures with large diameters (10 to 50 m). On the images, the structures project long acoustic shadows and are associated with significant variations in the depths of the first return. This indicates that they are steep-sided and elevated above the surrounding seafloor by at least 20 m. Characteristic of these features is a cauliform texture indicative of a hummocky surface. Samples taken from these features have shown them to be coral mounds, constructed by *Lophelia pertusa*. Their shape, size and texture is typical of living hemispherical colonies. They are formed by sub-units of up to 1.5 m in diameter. The average thickness of the mounds is 20 to 25 m and their diameter usually reaches 50 m. Figure 8.20 corresponds to a close-up view of one of these mounds. It is made of sponge-encrusted dead coral, covered by a thin layer of sediments. This accounts for its regular shape and the smooth acoustic appearance. On top of the mound lies a smaller structure, more reflective and with a shadow of several metres, formed of living corals. A linear structure crosses the entire image and is visible at the base of the shadow. It was found to be a submarine cable, lying directly on top of the two coral mounds. Juxtaposition of cables and hard, sharp outcrops such as coral, is often a cause of abrasion and eventual failure of these expensive and vital submarine installations.

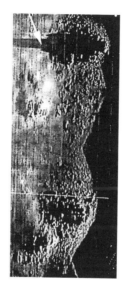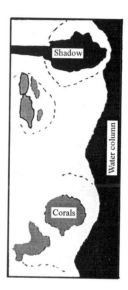

Figure 8.19. Unprocessed 500-kHz analogue record of coral mounds on Sula Ridge (Norway). Ensonification is from the right. The original image (left) has been inverted to show low backscatter as dark and high backscatter as bright (middle). The interpretation (right) outlines in grey the youngest coral colonies. The dashed lines indicate the approximate extent of the largest mounds. The black vertical lines are spaced 10 m apart. Courtesy of A. Freiwald, University Bremen, Germany.

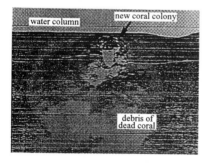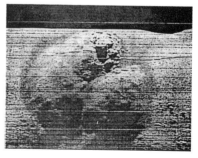

Figure 8.20. Unprocessed 500-kHz analogue record of a coral mound on Sula Ridge (Norway). Ensonification is from the top to the bottom of the image. The 50-m wide dead coral mound is covered by smooth sediments and capped by a young, living, coral colony. Courtesy of A. Freiwald, University Bremen, Germany.

8.7.2 Seafloor Vegetation

Seafloor vegetation is another manifestation of biological activity. Because it requires photosynthesis, it is restricted to the photoic zone (i.e. the zone where sunlight can penetrate) and hence relatively shallow waters. Other influences on the presence and types of seafloor vegetation are the composition of the substrate, the action of currents and waves, and the clarity of the water. Figure 8.21 shows a patch of seagrass (*Zostera marina*) on a sandy bottom. It was acquired at a frequency of 100 kHz, the sonar being towed 4 m above the seafloor in water depths of 10 m and less. Backscatter from the surrounding sand is low and homogeneous. The strong, localised echoes correspond to the seagrass. The thin, short leaves of the *Zostera marina* interact differently with the high-frequency acoustic waves, causing the rather heterogeneous appearance of this portion of the image.

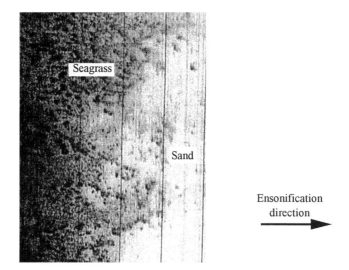

Figure 8.21. Unprocessed 100-kHz analogue record (EG&G Mk. 1) of seagrass on a sandy, shallow seafloor. Strong echoes are dark. The black vertical lines are spaced 10 m apart. Courtesy of the Geological Institute, University of Kiel, Germany.

8.7.3 Schools of Fish

Schools of fish are visible with sidescan sonar, especially at higher frequencies and in shallow waters. They can be identified because of the large acoustic contrasts of their gas-filled swim bladders. They apear as clusters of strongly backscattering points, whose shapes and dimensions are proportional to the number and density of the fish in the shoal, as well as the size of each fish. The image presented in Figure 8.22 was acquired in Eckernförde Bay (Baltic Sea). The 500-kHz sonar was towed about 15 m above the seafloor. The low backscatter seafloor in this area, crossed by a number of linear features, is interpreted as intensively trawled mud. Dark points (i.e. with a strong backscatter) are visible in the right portion of the image, and in the whole water column in the left portion. They correspond to shoals of mid-water herring (*Clupea harengus*). The exact dimension of these shoals are not straightforward to measure, as the fish are likely to move in all directions at an undertermined speed, and may have been imaged several times on the image. In the present case, they are not numerous enough to mask the structures on the ground.

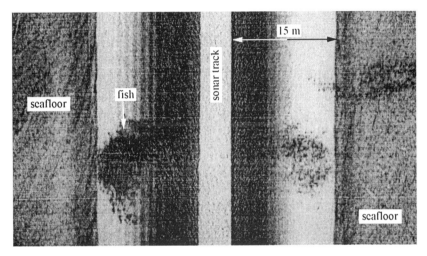

Figure 8.22. Unprocessed 500-kHz analogue record (Klein Model 520) of schools of fish over intensively trawled muddy sediment. Strong backscatter is dark, low backscatter is bright. Courtesy Geological Institute, University of Kiel, Germany.

8.8 ANTHROPOGENIC STRUCTURES

8.8.1 Anchor Tracks

Visible everywhere in fresh water and shallow marine environments, anchor tracks are particularly widespread in harbour areas. Generally localised, they are characterised by sharp, clearly defined furrows, that are highly visible on sidescan sonar imagery. The 100-kHz image presented in Figure 8.23 was acquired during the survey of a Baltic port in Germany. The sonar was towed 6 m above the seafloor, in an average water depth of 15 m. The disturbed seafloor, which shows evidence of intensive surface reworking in all directions, is interpreted as mud. Several elongated and narrow (< 2 m) marks are visible. Their edges are strongly defined, and the variations in backscatter show them to be depressions in the seafloor, less than a metre deep. These marks are the tracks left by ships' anchors. Discontinuous marks are caused by jumping of an anchor, and curved tracks by anchors dragging across the seafloor.

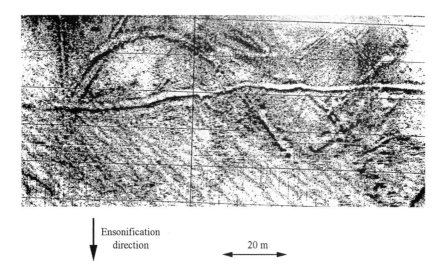

Ensonification
direction   20 m

Figure 8.23. Unprocessed 100-kHz analogue record (Klein Model 590) of anchor tracks on a muddy harbour floor. Strong backscatter is dark and low backscatter bright. Courtesy of the Geological-Paleontological Institute, University of Kiel, Germany.

8.8.2 Trawl Marks

Trawl fishing has been active in most coastal environments and on continental margins for many decades. Otter-board trawl fishing (also called ground-fishing) is the most commonly used technique. Otter-boards are heavy plates that are attached to the bottom of fishing nets to ensure they stay close to the seafloor. Weighing up to several hundred kilograms, these boards are dragged across the seafloor and leave distinct imprints. This type of operation greatly affects the benthic community, thoroughly mixing surface sediments to a depth of more than 20 cm (Krost et al., 1990). Beam-trawling (an alternative ground-fishing technique) is less damaging but still has a significant impact on the seafloor. Knowledge of the type and frequency of commercial fishing in an area, and recognition of the various marks caused by these activities, is essential to assess the degree of environmental disturbance caused by the commercial fishing industry. Sidescan sonar imagery forms a vital component in this monitoring.

Evidence for trawl fishing is present all over the deeper parts of the Baltic Sea (Werner et al., 1990). This relatively shallow coastal sea is typical of an intensively fished region where trawl tracks may merge into patterns covering the entire seafloor over several square kilometres. Single trawl tracks can remain nearly unchanged for several years. Their density was analysed with computer-assisted techniques (see Chapter 10). Trawl-track frequency increases with water depth and decreasing grain size. This is due to the stronger mechanical resistance of sand to the action of the otter-boards, and the increasing predominance of mud with greater water depth.

Figure 8.24 shows a 500-kHz sonar image of the most intensively trawled part of Eckernförde Bay. The sonar fish was towed 6 m above the seafloor in a water depth of 25 m. The image has a homogeneous, moderate backscatter indicative of a muddy seafloor. The various types of otter-board tracks appear as broad, sharp, linear troughs and herring-bone patterns. Variations in their shape and pattern depend on whether the boards ran smoothly through the upper sediments or jigged, jumped and hopped across the seafloor. The average width of the otter-board tracks is around 2 m.

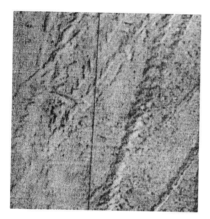 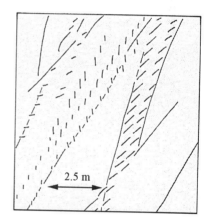

Figure 8.24. 500-kHz sidescan sonar image (Klein Model 520) showing intensively trawled muddy sediments with different types of otter-board marks left by ground-fishing activities. Strong backscatter is dark. The thin black line in the middle of the image is a calibration line and not a real feature. Courtesy of the Geological-Paleontological Institute, University of Kiel, Germany.

Sandy bottoms are stronger mechanically than muddy bottoms. Trawl tracks on such seafloors will be similar, but more poorly structured, than on muddy bottoms. At shallow depths, wave action tends to rework sand, slowly obliterating the traces of trawl fishing. Thus in non-cohesive sediments like sand, trawl marks tend to stay for only a few years, whereas in cohesive sediments like mud, they can remain almost unchanged for several years. Figure 8.25 was collected by a sonar towed 12 m above the seafloor, operating at a frequency of 100 kHz. The water depth is 20 m. The sandy seafloor returns a more heterogeneous backscatter than the muddy sediments of Figure 8.24. The seafloor here is criss-crossed by trawl marks, that are similar to, but less well defined, as those on Figure 8.24.

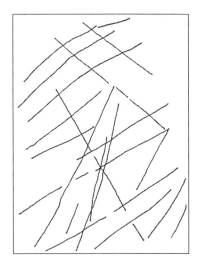

Figure 8.25. Unprocessed 100-kHz record of an intensively trawled area of sandy seafloor. Note the weak appearance of the otter-board trawl marks on this type of sediment. Strong backscatter is dark, low backscatter bright. Black horizontal lines are spaced 10 m apart. Courtesy of the Geological-Paleontological Institute, University of Kiel, Germany.

8.8.3 Wrecks

Shipwrecks are common occurrences in sidescan sonar images of coastal environments. They are more numerous close to the shores, where their visibility is enhanced by shallow water and hence lesser ranges to the sonars. In other environments, such as abyssal plains, shipwrecks are more difficult to find. Their appearances will range from a few reflective pixels to debris fields centred on the approximate location of the wreck (e.g. Parson et al., 1995). Wrecks reflect acoustic energy according to the present condition of the ship's constituent components, the extent and depth of burial in sediments, and the frequency of the sonar used to image them.

The first example was acquired at a frequency of 100 kHz (Figure 8.26). Located on almost homogeneous sandy mud, the outline of the wreck is still visible. Approximately 15 m long, it projects an important shadow (in white on this image) allowing its height to be determined as several metres above the seafloor. Oriented diagonally to the ship, a strongly reflective line (dark in this image) may be the mast of the vessel. The mottled area of low-reflectivity around the ship corresponds to a debris field. Diving identified the wreck as a wooden coal transport vessel.

Figure 8.27 shows another shipwreck, in a different environment, surveyed at a higher frequency (500 kHz). It was found in the North Sea, off Sylt Island (Germany). The sonar was towed 4 m above the seafloor in a water depth of 15 m. The substrate presents a moderate backscatter, indicative of mixed mud and sand. The vessel is at the junction of two ripple fields. The one in the top of the image exhibits a wider crest spacing. The one surrounding the wreck shows smaller structures, and seems related to current activity around the wreck. Strong echoes outline the vessel's skeletal framework, but the stern is partially buried by sediment. The elongated shape of the structure may be partly due to the absence of slant-range correction for this image.

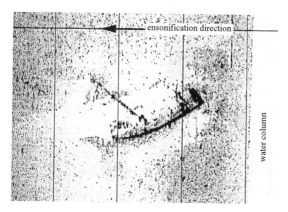

Figure 8.26. Unprocessed 100-kHz analogue record of a shipwreck in very shallow water. This wreck was identifed by divers as a wooden coal transport vessel, lying on a sandy seafloor. Ensonification is from right to left. Black vertical lines are spaced 10 m apart. Strong backscatter is dark, low backscatter bright. Courtesy of the Geological-Paleontological Institute, University of Kiel, Germany.

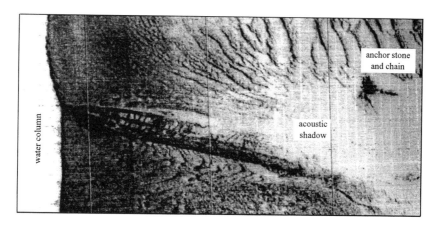

Figure 8.27. Unprocessed 500-kHz sonar record (Klein Model 590), showing a partially buried shipwreck lying on a sandy bottom in the North Sea. Black vertical lines are spaced 5 m apart. Strong backscatter is dark, low backscatter bright. Courtesy of the Geological-Paleontological Institute, University of Kiel, Germany.

Again, but with a different frequency and in a different environment, Figure 8.28 shows a shipwreck closer to shore. Surveyed at a frequency of 325 kHz, this shipwreck lies in very shallow water. The original image (left) was inverted to show strong backscatter as bright and low backscatter (including shadows) as dark. The local geology is evenly divided between homogeneous sand (left half) and a rougher seafloor (right half), made of cobbles. The wreck lies at the transition zone between the high-energy, wave-affected region near the beach, and a lower-energy environment in slightly deeper water which is less influenced by waves and currents. The brightly reflecting surface on the lower middle of the image is the ship's prow, and the thin linear reflectors are various ribs and bulk-heads. Its length is difficult to measure accurately, but is close to 25 metres. The poor condition of the wreck is largely a result of destruction by breaking waves.

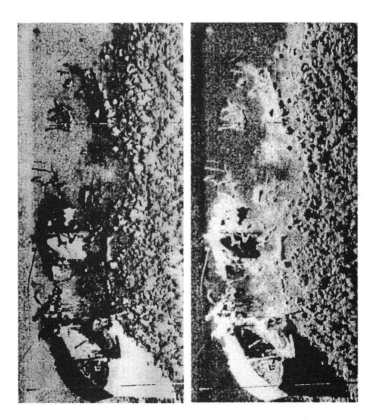

Figure 8.28. Unprocessed 325-kHz sidescan record (Dowty Widescan 3050) of a wooden wreck in the transition zone close to a shingle beach. The original image (left) was inverted to show strong backscatter as bright and low backscatter (including shadows) as dark. This mode of representation also enhances the visibility of the different elements of the shipwreck. Ensonification is from left to right. Courtesy of T. LeBas, Southampton Oceanography Centre, UK.

8.8.4 Dump Sites

Dump sites, whether organised or not, are another characteric trait of anthropogenic activity in coastal environments. Sidescan sonar was used in recent well-publicised surveys to locate and salvage barrels of toxic material dumped off-shore during the last few decades. High-resolution sonar imagery is likely to play an important role in monitoring various munition dump sites left scattered around European coasts after the Second World War. The objects disposed of are generally poorly sedimented, and clearly distinguishable from their background. However, the majority of dump sites visible in coastal environments are related to engineering tasks: construction of oil platforms, disposal of harbour mud after dredging, etc. The dumped materials are sand or mud, and are often similar in composition to the seafloor on which they lay.

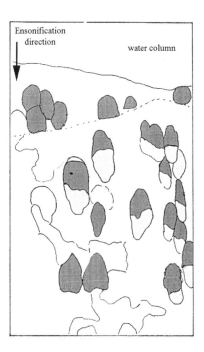

Figure 8.29. Unprocessed 500-kHz record (Klein Model 520) of a dump site in the central part of Eckernförde Bay. It is made of sandy material on a muddy bacground. Strong backscatter is bright, low backscatter dark. Acoustic shadows are stretched because of poor geometric corrections. White horizontal lines are spaced 12.5 m apart. Courtesy of the Geological-Paleontological Institute, University of Kiel.

Figure 8.29 shows a classical dump site of sandy material on a muddy seafloor. Located in the Baltic Sea, the dump site contains material excavated during the building of oil platforms and harbour extensions from the nearby coast. The image was acquired at a frequency of 500 kHz, and shows some across-track stretching due to the absence of geometric correction (see Chapter 3: Sonar Data Processing, and Chapter 9: Image Anomalies and Sonar System Artefacts). The surrounding seafloor shows a lower backscatter than the individual structures, indicative of mud. Individual dumps are actually circular and have dimensions of a few metres at most. The strong first returns on this image are an artefact caused by cross-talk between sidescan channels.

8.8.5 Other Anthropogenic Features

Since the end of the last glaciations, the shape and location of coastlines have altered significantly. Many archaeological sites once on dry land are now under water, close to shore (e.g. Ozette Point in Washington, USA), in river deltas (e.g. Alexandria, Egypt) or further into the sea. Figure 8.30 shows one such site, in a fjord of Northern Germany. It was collected with a 500-kHz sidescan sonar, towed in water depths of 20 m. The homogeneous backscatter of the image is indicative of a sandy seafloor. Two very distinct circles are visible, with mottled textures and diameters of about 25 metres. They comprise smaller rocks as well as a number of larger stones, arranged in a partially circular shape. These circles may represent Neolithic remains. Such observations of stone circles with sidescan sonar are relatively common in western Europe, where they date from the end of the last glaciation. Some linear marks from otter-board trawling are visible close to the stone circles.

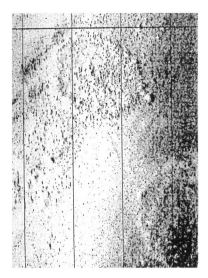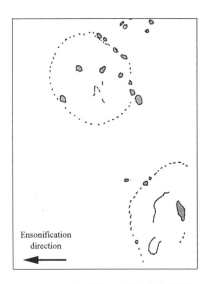

Ensonification
direction

Figure 8.30. Unprocessed 500-kHz sonar record (Klein Model 520), showing stone circles, of possibly Neolithic origin, from a German fjord. Strong backscatter is dark, low backscatter bright. The black vertical lines are spaced 10 m apart. Courtesy of the Geological-Paleontological Institute, University of Kiel, Germany.

8.9 SUMMARY

The structures observed with sidescan sonar in coastal environments are very distinct from those observed in other environments. The combination of the shallow water depths, land influence and anthropogenic activity, all concur to make the coastal seafloor an important and distinctive place. The conditions of data acquisition are very different in coastal environments compared with deeper waters. Sidescan sonars are generally towed much closer to the seafloor than is safe or possible in other, deeper, environments and often at much higher speeds. The small scale at which many, important, coastal seafloor features are visible requires the use of higher acoustic frequencies. Because of the proximity to shore, and to identifiable landmarks, navigation of coastal sidescan sonar surveys is often more stringent than in the deep sea. The large number of surveys performed in coastal waters, with generally specific objectives and of short duration, combined with their limited budgets, often means that the processing of such sonar imagery remains in low demand.

In this chapter, we have endeavoured to present examples of the diverse coastal processes and their implications. The distribution of bedforms such as ripples, for example, is a good indicator of the hydrodynamic conditions. Sidescan sonar imagery is an excellent tool for analysing and monitoring the changes, natural and anthropogenic, undergone by the environment. With the increasing use of coastal areas, and with the possibilities of global warming and its associated sealevel changes, such information is a prerequisite to the successful protection of seafloor and coastal installations.

8.10 FURTHER READING

- **About sedimentary features:**

Reading, H.G.; "Sedimentary Environments and Facies", Blackwell Scientific Publications: Oxford, 615 pp., 1978

Reineck, H.-E., I.B. Singh,; "Depositional Sedimentary Environments", Springer Verlag, New York-Heidelberg-Berlin, 551 pp., 1980

Elverhoi, A.; "Glacigenic and associated marin sediments in the Weddell Sea, fjords of Spitsbergen and the Barents Sea: a review". Marine Geology, vol. 57, p. 53-88, 1984

Hambrey, M.; "Glacial environments", UCL Press Limited, London, 296 pp., 1994

- **About pockmarks:**

Hovland, M., A.G. Judd, A.G.; "Seabed pockmarks and seepages- Impact on Geology, Biology and the Marine Environment". Graham and Trotman Limited, London, 293 pp., 1988

- **About biological activity:**

Newton, R.S., A. Stefanon; "Application of Sidescan Sonar in Marine Biology". Marine Biology, vol. 31, p. 287-291, 1975

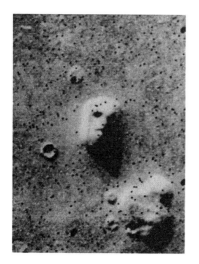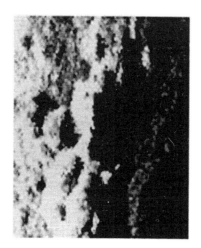

Figure 9.1. The "Face on Mars" (Viking imagery, copyright NASA-JPL) and the "Face on the Seafloor" (TOBI imagery, copyright SOC), two examples of image artefacts created by the conditions of acquisition and liable to misinterpretation.

9

Image Anomalies and Sonar System Artefacts

9.1 INTRODUCTION

The previous chapters have demonstrated all the stages of data acquisition and interpretation, in all depths and marine geological environments so far studied with sidescan sonar. Sidescan sonar imagery, like any data, is rarely devoid of anomalies and artefacts. They may be easy to spot or mistaken for real features, and they may be difficult to interpret or remedy. The present chapter aims at showing all possible sources of errors and artefacts, how they can be avoided during the processing, and how to recognise and interpret them when they occur. This will be demonstrated by drawing both on the most recent theoretical studies on the subject, and on real-world examples from a variety of applications. The interested reader may also find it profitable to look at a short publication from EG&G Marine Instruments (Fish and Carr, 1990) which presents shallow-water sidescan sonar operations, mainly for the detection of man-made structures. The different sections of this chapter follow the acoustic wave from transmission to reception and processing. This includes propagation through the water column, backscattering toward the sonar platform, processing and the final interpretation. All these stages are prone to errors and artefacts; some of them are unavoidable, but all of them are recognisable.

9.2 WATER COLUMN ARTEFACTS

Artefacts related to the propagation of the acoustic waves in the water column from the platform to the seafloor and back can be attributed to two sources. The first are variations in the structure of water column itself. These can be density variations, salinity variations or temperature variations. They arise from the stratification of the ocean and can be complicated by horizontal variation and, very frequently, by fluctuations in the stratification (i.e. internal waves). The importance of these effects will differ according to the type of sonar used: deep-towed or shallow-towed, short-range or long-range. For deep-tow sonars (e.g. TOBI, DSL-120), the water is almost isothermal and acoustically fairly uniform. Hence the propagation effects are constant and predictable (Somers, 1993). Shallow-towed sonars (eg GLORIA) will make use of most of the water column. Depending on the depth, a certain amount of thermocline layers will modulate the depth and angle at which the acoustic rays propagate. They can give rise to patterns at the far range that are very reminiscent of the linear bedforms associated with a bottom current, and which can only be treated by cutting off the outer portion of the image (Somers, 1993). Shallow-towed sonars in shallow waters (e.g. EG&G-272T) should be less prone to these variations. However, there have been reports of shoal-like structures produced by temperature inversions in summer, with warm-over-cold layers bending the sonar rays up (Fish and Carr, 1990). Other causes of local variations in the properties of the water column can be cold and hot water seeps (anthropogenic or not), and lateral variations in salinity (e.g. near large river estuaries).

Sizeable heterogeneities of the water column can be produced by the presence of bubbles. They may come from the wake of the survey ship (or neighbouring ships) or from the cavitation caused by the ship's propellers. These bubbles have lifetimes measurable sometimes in hours. High-frequency systems are quite sensitive to the longer-lived, very small bubbles. The sonar beams become partially dispersed and partially reflected before they reach the seafloor, creating random data gaps at all ranges. The influence of hydrothermal vents (see Chapter 5: Mid-Ocean Ridge Environments) on the propagation of acoustic beams is a different matter, as their flows are usually enriched with minerals. Interference has been documented in some instances with shallow-towed lower-frequency sonars (about 12 kHz), but not, to our knowledge, with higher frequencies. This may be due to the size of the bubbles, and to their acoustic contrasts with the surrounding water.

Heterogeneous propagation of the sonar beams through the water column can also be affected by biological scattering. Well-dispersed plankton in the Deep-Scattering Layer may affect deep-water shallow-towed sonars such as GLORIA, but is usually not directly visible on sidescan images. The scattering strength of fish is considerably increased by their possession of a swim-bladder full of gas. Depending on sonar frequency, fish can appear as strong discrete targets, the size of which varies with the number and dimensions of the fish involved (see Figure 8.21). In extreme cases, whales and dolphins have been known to try to communicate with a sonar by transmitting at the same frequency. The resulting structures are usually highly reflective streaks elongated along-track.

9.3 RADIOMETRIC ARTEFACTS

The first and most frequent cause of systematic radiometric artefacts resides in the acoustic system itself. The sonar cable is a fragile link between the platform and the recording and processing systems aboard the ship. There are different kinds of cables, depending on the depth and transmit-rate requirements: reinforced lightweight cables for shallow operations (e.g. EG&G-272T), armoured cables for deep-towed platforms (e.g. TOBI), fibre-optics for high transmission rates (e.g. DSL-120). Lightweight cables have a waterproof coating, but the primary protection, as for most of the other cables, is the conductor insulation itself. The immersed part of the cable ("wet" cable) is very vulnerable to any unseen obstacle: underwater moorings or fishing gear, and even fish bites. The connections, or terminations, at one end of the cable or the other are likely causes of problem as well. Faulty or broken contacts, will cause any kind of radiometric artefacts, from the occasional loss of data to a total "black-out" of the system. Other sources are related to the platform's design (e.g. cross-talk between badly placed transmitting arrays, electronic devices with close electromagnetic frequencies interfering with the sonar's electronics). They are usually resolved before operational surveys, or can be checked before each deployment (e.g. watertight integrity of the electronics).

Another cause of radiometric artefacts is the interference with other acoustic sources. Passing ships are discrete sources of acoustic noise, and affect the lower-frequency, long-range systems particularly. For example, a supertanker radiates enough noise to be very detectable on a GLORIA image to ranges of over 10 kilometres. Higher-frequency systems may also be affected when surveying busy shipping routes. The effect of neighbouring vessels is a cross-track band of diffuse noise. It can be suppressed by locally reducing the TVG level close to the nadir (Somers, 1993). A similar phenomenon occurs with icebergs, which are a prolific source of noise and also reflect large amounts of energy. They appear as discrete targets, reinforced by the pronounced surface duct produced by the strong surface cooling. Although icebergs do not cast acoustic shadows, they might still be mistaken for objects on the seafloor. The last, and most common, acoustic interference is created by instruments run concurrently, on the survey vessel or on neighbouring ships: depth profilers, seismics, etc. They create continuous interference patterns, which are not always easy to remove. The interference can be reduced by increasing the distance between the conflicting instruments, when feasible, or ensuring they operate at different, non-harmonic frequencies.

The type of radiometric artefact common to all sonars is the irregular band of high-backscatter points nearest to the ship's track (Figure 9.2). This band is visible on all sonar images presented in this book. It is explained by contribution from near-normal incidence specular and sub-bottom returns, which will only occur within the first few degrees from nadir. Unfortunately, there is no way of directly removing this artefact short of masking it completely. An alternative approach is to create a synthetic backscatter image from swath bathymetry immediately beneath the vehicle's track (Somers, 1993). This approach gives a more continuous image, but is limited to lower-resolution sonars like GLORIA.

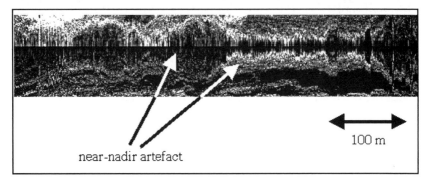

Figure 9.2. Irregular bands of high-backscatter points close to the ship's track are a systematic artefact on all sidescan sonar images.

The across-track attenuation of the backscattered waves with increasing distance between the seafloor and the sonar platform is corrected with angle-varying gain (see Chapter 3: Sonar Data Processing). Non-systematic variations in the beam pattern and the angle of incidence over a uniform seafloor may not be corrected by AVG. The structures imaged in other regions will appear as lower or brighter patches at the same range along the image. They are usually not difficult to interpret visually, because of the continuity with neighbouring structures. Problems may arise when using digital methods which use the grey level statistics (see Chapter 10: Computer-Assisted Interpretation). Correcting algorithms are fortunately available in the literature (e.g. Reed and Hussong, 1989; LeBas et al., 1994). To use them effectively, it is important to know what AVG was originally applied to the sidescan sonar data.

Another possible radiometric artefact is the rapid attenuation of the backscattered signal when the sonar platform goes up suddenly (when hauled back, for example to avoid an obstacle, or if there is a rapid change in the seafloor's depth). This change is usually too localised and rapid to be corrected with the normal time-varying gain (Figure 9.3), and an appropriate TVG needs to be specifically computed. If the depth change is too important, the sonar receiving times will fit no longer, and no more data will be recorded.

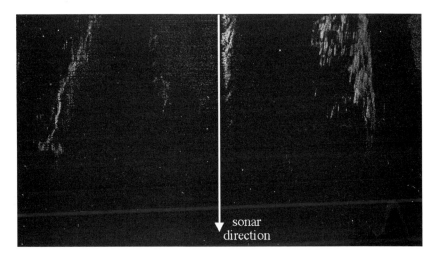

Figure 9.3. Example of sonar imagery acquired when the vehicle is hauled back in a straight line. In less than a kilometre (the length of the image), reflections from the seafloor are attenuated and then disappear. The most reflective and/or higher structures are visible longer.

9.4 GEOMETRIC ARTEFACTS

9.4.1 Variations in Survey Speed

Sidescan sonar data is distorted by the variations in the motion of the towfish. Most, if not all of these variations are accounted for during processing (Chapter 3: Sonar Data Processing). They are visible if the sonar image has not been fully corrected, or has been badly corrected. The first factor is related to the speed of the survey vessel. Along with the pulse-repetition frequency (Chapter 2: Sonar Data Acquisition), this speed dictates the distance between each swath line. If variations of the ship's speed are not correctly taken into account, the image will be distorted in the along-track direction (Figure 9.4). If the platform speed assumed during processing is higher than the actual value, the swath lines will be positioned too far away from each other, and the image will be stretched along-track. Conversely, if the platform speed is lower, the swath lines will be positioned too close to each other, and the image will be compressed along-track. In a recent survey in which one of us (Ph. B.) participated, the speed assumed for processing was 1.5 knot (2.8 km/h). In fact, because of winds and surface currents, the real ship's speed was varying between 1.2 and 2 knots (2.2 km/h and 3.6 km/h). The survey lines were up to 15 km long, and processed images were showing discrepancies as large as 100 m between matching features (such as fault scarps) on adjacent lines. This prompted the reprocessing, and the reduction of the mismatches to a few metres (i.e. close to the limit of resolution of the sonar). It is also important to remember that large variations of the towing speed will cause the sonar platform to pitch either down (when slowing) or up (when accelerating).

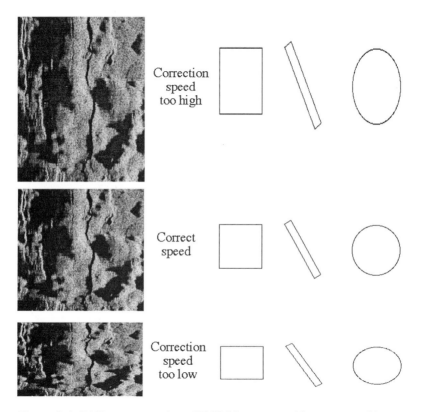

Figure 9.4. Different examples of TOBI imagery, with correct and incorrect speed corrections. The line features at right show the along-track deformation of stylised shapes.

9.4.2 Variations in the Platform's Altitude - Heave

Variations in the altitude of the sonar platform are generally accounted for by using a towfish-based altimeter or computing the length of cable out. These methods are not always precise, and some unaccounted fluctuations may occur. If the altimeter's frequency is relatively low (e.g. 3.5 kHz), the signal can penetrate bottom sediments before encountering a reflector hard enough to return the signal. This would cause the altimetre to see the seafloor as deeper than it appears to the sidescan sonar. Conversely, a signal at a higher frequency may be reflected by extremely soft or fluid sediments, or local heterogeneities in the water column. This produces an early return, which shows the seafloor shallower than it really is. In some cases, the altimeter "locks" onto these false echoes and gives erroneous depths for an appreciable time (i.e. along-track distance).

The movement of the sonar platform around its nominal altitude above the seafloor is called the heave. If the distance to the seafloor is larger than it should (heave > 0), the sonar will record backscatters from a larger swath than usual (Figure 9.5). But the processing algorithm will assume the swath is the same, and it will place the seafloor reflections closer to the nadir than they really are. Conversely, if the seafloor is closer than it should (heave < 0), the recorded swath will be narrower (Figure 9.5). The processing algorithm will place seafloor structures further away from the nadir than they should be. Both these processes create across-track distortions of the same amount on both sides.

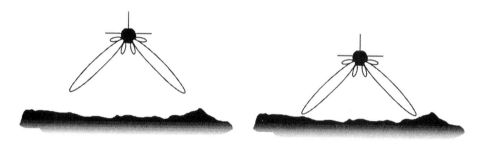

Figure 9.5. When the sonar is higher above the bottom, the swath width will increase. If the system erroneously records higher altitudes than the real one, the image will be compressed across-track. Conversely, if the system records lower altitudes, the image will be stretched across-track.

These across-track variations are symmetric and therefore easier to pick up. Figure 9.6 shows a partially processed TOBI image in a mid-oceanic ridge terrain. Aligned along-track, a large axial volcanic ridge is visible in the rightmost portion of the image. Two long white lines span the entire image. They correspond to sea surface reflections (see section 9.6) and should be nearly parallel to the sonar track. In fact, they show as arcuate features moving away from the nadir. The structures on each side, and particularly the summit line of the volcanic ridge and the fault scarp at the extreme left, are similarly curved. These symmetric variations are as large as 600 metres, and are attributed to unprocessed variations in the platform's altitude above the seafloor.

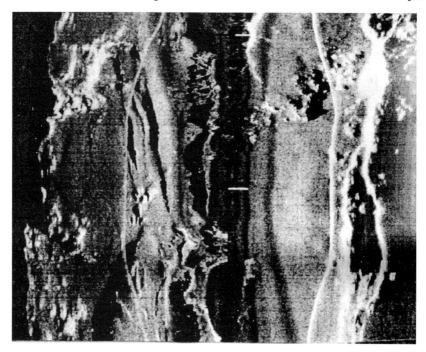

Figure 9.6. Unprocessed variations in the sonar's altitude. This TOBI image is 6 km wide. The sea surface reflections are deviating from straight lines by as much as 600 metres.

9.4.3 Unprocessed Roll

Roll is the movement of the sonar platform around its longitudinal axis. One transducer will point higher up than intended, and the one on the other side will point lower down (Figure 9.7). Rolling may result in intensity distortions in the area immediately below the towfish because of the rotation of the vertical sidelobes. The transducer pointed up will image structures at a larger range, but the processing system will compress them down to the assumed swath width. Conversely, the transducer pointed down will image at a shorter range than usual, and objects on the seafloor will be stretched by the processing system to accommodate the full swath width. The distortions created by unprocessed roll will therefore occur across-track, and be consistently asymmetric. If two elongated structures are present on the seafloor, the one on the upturned side of the sonar will be positioned closer to the sonar track than it really is, and the one on the downturned side will be positioned further away from the track.

A sonar towed at 100 m above the seafloor, and affected by an unprocessed roll of 10°, will misposition features along-track by 17.6 metres (and more if the height above the seafloor increases). If unprocessed roll decreases to 1° (the limit of resolution of most attitude sensors), the mispositioning will decrease to 1.74 metre, which is much more acceptable for most surveys.

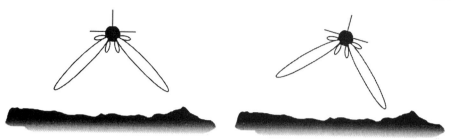

Figure 9.7. Unprocessed roll of the sonar platform creates asymmetric across-track distortions of the image.

9.4.4 Unprocessed Pitch

Pitch is the orientation of the sonar platform around the horizontal plane (Figure 9.8). When the towfish tilts up (pitch > 0), it looks further away than intended. The sonar will record a swath line that it will resurvey later. Conversely, when the sonar tilts down (pitch < 0), it will replicate a swath line previously surveyed. Brutal and unaccounted changes in pitch will result in structures on the seafloor being suddenly interrupted at all ranges, and resuming after a few swaths (if the pitch goes back to normal). Pitch is usually associated with changes in the towfish's altitude or its speed.

Figure 9.8. Unprocessed pitch of the sonar platform will replicate previous swath lines (pitch < 0) or anticipate future swath lines (pitch > 0).

9.4.5 Unprocessed Yaw

Yaw corresponds to a side-to-side movement of the sonar platform around its tow path. The swath lines are rotated along the central axis of the platform (Figure 9.9), creating along-track and across-track distortions. The half-swath inside the turn will reimage some features already imaged, but from a different look-angle. The half-swath outside the turn will image some features further away than intended, and from a different look-angle.

Yaw-like movements in wide swaths can occur when long cable lengths and depressors are used. These movements are referred to as "kiting", and comes from damaged or unproperly attached depressors. Under the effects of bottom currents, or because of problems with the tail fins, the sonar may also exhibit a permanent yaw in one direction ("crabbing").

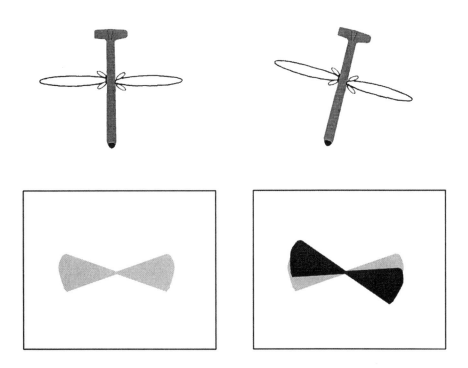

Figure 9.9. Unprocessed yaw of the sonar platform (top) creates asymmetric across-track and along-track distortions of the image (bottom).

The variations of altitude, roll, pitch, and yaw are usually simultaneous. The platform's speed, course and orientation will vary under the influence of underwater currents, changes in the sonar-survey ship geometry induced by wind and sea currents or (in the absence of a depressor weight) caused by movements of the ship communicated to the towfish through the cable. For example, at the end of the survey line, and to start a parallel line in the other direction, the ship will make a 180° turn. If the cable length stays identical, variations in the ship's speed and course will make the cable slacken; the deep-tow sonar will first sink, and then turn. To maintain a constant, safe, altitude above the seafloor, the sonar is generally hauled back a little during the turn, which augments slightly the altitude at the beginning of the turn. Both options create variations in the altitude, the roll (turning more or less in the direction of the turn), the pitch (the sonar points upward at varying angles, depending on the speed at which it is hauled back), and the yaw (the sonar oscillates quickly around its central axis). State-of-the-art attitude sensors can now measure relative movements with accuracies smaller than 0.1°, and processing software can correct the images accordingly (see Chapter 3: Sonar Data Processing).

9.5 PROCESSING AND OUTPUT ARTEFACTS

9.5.1 Beam Spreading

Acoustic waves propagating in the ocean are subject to spherical spreading. The principal effect will be the widening of the beam width with distance from the sonar, and the subsequent decrease in the along-track resolving power (Figure 9.10). In the near-nadir regions, two close targets will be resolved by the sonar. At further ranges, two identical targets at the same distance from each other will be imaged by the same beam, and appear as one single target.

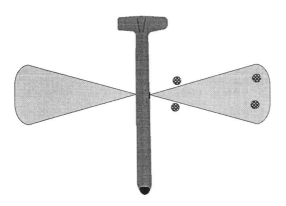

Figure 9.10. Exaggerated representation of beam spreading. In the near-nadir regions, the beam is narrow enough, for two closely spaced targets to be imaged separately. Further away, identical targets will be imaged with the same beam and blended into one single, larger, target.

Beam spreading constantly degrades the along-track resolution at far ranges. For example, the along-track resolution of GLORIA is 45 m for a pulse-repetition frequency of 0.033 s^{-1} (1 ping every 30 s). At a ship speed of 8 knots (14.8 km/h), the along-track resolution is 120 m near the nadir. Because of beam spreading, the along-track direction increases to 900 m at a 22-km range. Advanced sonar processing techniques (e.g. Mason et al., 1992) use the knowledge of the point-spread function of the sonar system in use (i.e. the actual pattern of the beam) and constrained iterative restoration algorithms such as the Jansson-van Cittert method.

If beam spreading is important, and corrections are not applied or are incorrectly performed, the discrete targets in the far range will be elongated along-track. They will appear longer than they really are, and individual targets will be erroneously merged (Figure 9.11).

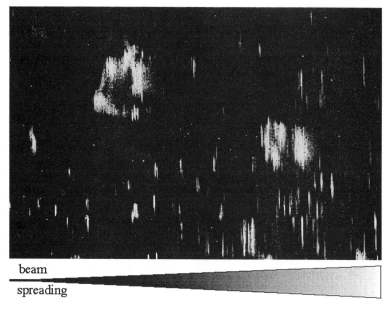

Figure 9.11. High-resolution sidescan sonar image showing the effects at far range of uncorrected beam spreading.

9.5.2 Slant-Range Corrections - Layover

Slant-range corrections remap individual pixels from their apparent positions across-track to their true positions (see section 3.4.1). It assumes a plane seafloor across-track. This can be a problem in areas of high relief, or above regional slopes (Figure 9.12). The first echo to return to the sonar platform will be from point A on the seafloor (i.e. closest to the sonar). This point is higher on the ground than the true nadir, and at some distance from it. Performing a slant-range correction with a flat bottom assumption will incorrectly assume A is the nadir, and move the central line of the processed image to point B (the real nadir). When the target's echo arrives at the sonar, it is assumed to be at the same depth as the original return. The slant-range correction places it in the image at point D, rather than at its actual location (point C).

For portions of the seafloor sloping up from the nadir of the sidescan fish, the target will appear closer to the nadir than it really is. In particular, topographic features will be positioned systematically closer; they appear to be foreshortened, or to "lean" toward the nadir. A distant peak would be located before its flanks. This phenomenon is known as layover. Where the bottom slopes away (down) from the nadir, the target will appear further away than it really is. These distortions can be important: examples with SeaMARC II imagery show displacements of up to 1,000 m (for an across-track resolution of 10 m), and an apparent rotation of 20° of fault scarps on slopes (Reed and Hussong, 1989). In mosaics containing parallel tracks with opposite look-angles, features will be displaced in opposite directions.

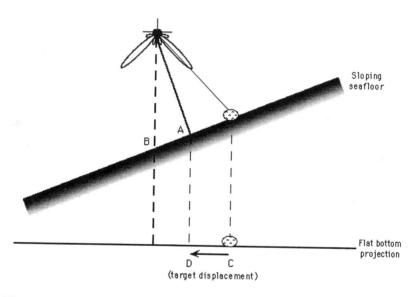

Figure 9.12. Illustration of the layover effect. Structures on the slope facing toward the sonar will appear closer than they really are; structures on the slope facing down from the sonar will appear to be further away than they are in reality.

Correction of layover and foreshortening effects is simple in theory, and lies in determining the actual depth of the target away from the track line. This requires the determination of the seafloor slope and the height of the fish above the bottom. Co-registered bathymetry, acquired from phase information or from other sources, is a prerequisite for accurate slant-range correction.

9.5.3 Processing Artefacts

Image artefacts can be created at different stages in the processing if incorrect parameters are chosen, or if some correction stages are omitted. Pre-processing the navigation data (section 3.2.2) ensures that the positions of the survey vessel and of the sonar platform with respect to the ship are accurate. Incorrect calculations will show in mosaics, when corresponding structures on adjacent lines are offset or exhibit angles rapidly changing at the intersection of the swaths.

Pre-processing the attitude data (section 3.2.3) ensures that outliers and spikes, or long-time offsets with no real significance are removed. If some values subsist, they will create local geometric distortions attributable to the variations in heave, roll, pitch and yaw, and whose effects have been shown in the previous section.

The requantisation of individual backscatter values (section 3.3.1) usually does not create problems. Inappropriate requantisation schemes will create flattened radiometric ranges, with small contrasts. Other radiometric corrections include time-varying gain and angle-varying gain. Inappropriate TVG gains will show as series of swaths darker or lighter than adjacent swaths (Figure 9.13). Inappropriate AVG gains will show as darker or lighter patches aligned along-track at definite range intervals.

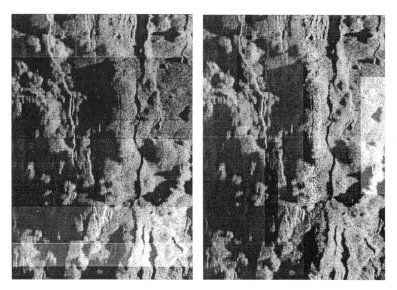

Figure 9.13. Examples of inappropriate time-varying gains (left) and angle-varying gains (right) in TOBI imagery.

Radiometric problems leading to across-track striping have been explained in section 3.3.3. Sometimes, swath lines appear "shifted" across-track. Most of the swath line is black, and the far-range portion shows features which should be at near-range. This is explained by the sonar's altimeter losing the bottom and subsequently timing the backscatter returns inaccurately. One example was visible in Chapter 7 (Continental Margins), on Figures 7.6 and 7.7 (Veatch Canyon). The portion at the end of the received beam can be matched back to fit the rest of the canyon (Figure 9.14).

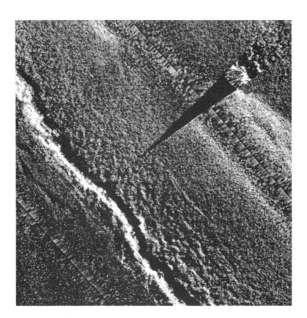

Figure 9.14. Example of an across-track artefact in GLORIA imagery.

9.5.4 Output Artefacts

Often neglected, some artefacts do not come from the original data, but from the way in which it is displayed and presented for interpretation. Up to the beginning of the 1980s, almost all sidescan sonar imagery was made available in near real-time by printing on electrosensitive paper. This is still the case for small-scale, low-cost surveys and as a backup during the other surveys. The themes to remember when looking at paper records is that all radiometric and geometric corrections will not necessarily have been performed. For example, slant-range may be corrected (assuming a flat seafloor, evidently). But the anamorphosis will be performed with a constant ship speed fixed by the operator, and not always reflecting the actual speed. Structures may be elongated or compressed along-track. This is particularly noticeable when attempting to mosaic paper records of adjacent survey lines: the structures do not always match well, and are sometimes offset by hundreds of metres. Because of the finite size of the paper, geometric corrections taking into account the attitude of the sonar platform are generally not performed; small turns will be projected on the same line, artificially curving linear structures such as fault scarps. There are also risks of saturation or under-saturation, as the dynamic range will differ according to the quality of the printer. When printing for hours at a time, some types of printers will add streaks of artificially uniform data, although the digitally recorded data is fine. Finally, old paper records will not stay constant with time. Some of them will fade, changing the apparent radiometric response of terrains; others will crumple slightly, mottling the small-scale textures. All these problems seem trivial. They are, however, routinely encountered, and should always be kept in mind when attempting interpretation of paper records.

9.6 INTERPRETATION ARTEFACTS

9.6.1 Subsurface Reflections

Studies of deep-sea sediments show that they often have near-surface sound velocities very close to the sound velocity in the overlying water (a few per cent lower). It is therefore not unnatural to observe volume reverberation. The incident acoustic wave penetrates into the sediments and is backscattered by subsurface reflectors. Significant deep-sea sediment penetration occurs at frequencies of 12 kHz or lower ($\lambda_{inc} > 12.5$ cm). The depth of acoustic penetration depends on the frequency of the sound and the physical properties of the sediments involved. It can vary between several metres with GLORIA (6.5 kHz) and a few centimetres with DSL-120 (120 kHz).

The buried features scatter more strongly than the sediment interface and are therefore visible in the images. An example was seen in Chapter 6 (Abyssal Plains and Basins) on GLORIA imagery close to Hawaii (Figures 6.19 and 6.20). Huge fields of lava flows were buried under several metres of sediments, but were still perfectly visible on the image. Additional information (profiler data, optical images) is necessary to demonstrate with certitude the presence of subsurface penetration. This type of effect is not widespread, but should nonetheless be kept in mind during the interpretation of imagery acquired in sedimentary areas. Subsurface penetration is known to occur in radar imagery as well, mainly in very dry regions (Elachi et al. (1984) were able to detect fossil fluviatile channels 15 m below the surface of a desert).

9.6.2 Interference Effects

Another artefact is sometimes observed over flat sedimented seabeds. The sound waves are normally backscattered at the sediment-water interface, but a small proportion penetrates below the seafloor and is scattered by subsurface reflectors. The two acoustic waves interfere, constructively or destructively, to create interference fringes. Their geometry is similar to that of optical fringes caused by a thin oil film on water or of caustics caused by diffraction of a laser beam. These interference fringes require a transparent sediment layer and a good reflecting layer in close vertical proximity. If the subsurface layer is too deep, the differences in amplitude between the interfering waves is not high enough to create interference patterns. The optimal depth of the second reflector needs to be a whole or a half multiple of the sonar's wavelength for interference fringes to occur.

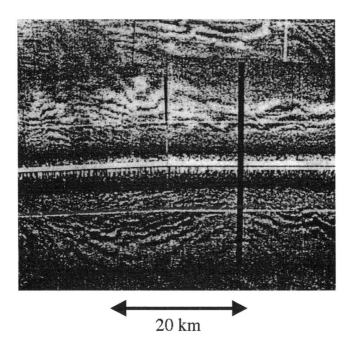

20 km

Figure 9.15. Hand-made mosaic of GLORIA paper records acquired in the Bering Sea, showing interference fringes. The sonar track is the white line in the middle. The high backscatter bands increase in width with the range from the sonar track. Widths are varying from 100 m at near range (< 5 km) to 2,000 m at far range (22 km).

Such fringes were first recognised during the survey of the US Exclusive Economic Zone in the NW Bering Sea (Huggett et al., 1992). GLORIA images were exhibiting strong interference patterns, sub-parallel to the ship's track and 100 to 2,000 metres wide (Figure 9.15). These interference patterns were always sub-parallel to the ship's track, regardless of heading and course changes. If the patterns were related to subsurface structures, they would be expected to remain similar with varying look-angles. Repeated surveys showed the persistence of these patterns over time, and that they were not related to sea conditions (unlike fringes created by the Lloyd's effect, see below). Since then, interference fringes have been observed throughout the world, at quite different frequencies, with SeaMARC-II and with TOBI (Figure 9.16). There seem to be no common factors between the geological characters of the sediments that show interference fringes. The only requirement is that the right acoustic layer lies on a reflective subsurface layer at the depth appropriate for the frequency of the system.

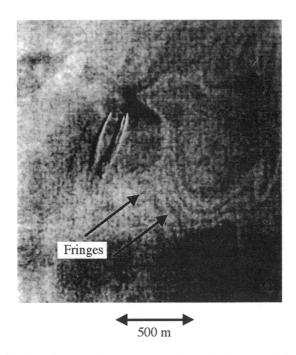

500 m

Figure 9.16. Interference fringes seen with TOBI in the Madeira Abyssal Plain . They were apparently formed within a first layer of sediments 25 cm thick. Courtesy D. Masson, SOC (UK.)

Interference fringes are not always attributable to rescattering of acoustic waves inside the sediment layers. A local interaction, which sometimes occurs at the surface, is the Lloyd's mirror effect, arising from interference between the returning bottom reverberation and its image on the sea surface. To be observed, this effect requires a high degree of coherence in the energy reflected from the sea surface at small grazing angles, and a transducer depth which does not exceed a few tens of wavelengths (Somers, 1993). Otherwise, the fringes are too closely spaced to be resolved. Since the fringes always present themselves as lines parallel to the track, they require identical reflection conditions over several pulses, which implies very close to a flat calm sea. This effect has been mainly observed with long-range shallow-towed sonars such as GLORIA. Deep-towed systems cannot generate Lloyd's mirror fringes because the path to the surface is too great.

9.6.3 Multiple Reflections

The acoustic wave transmitted by a sonar has three possible ways to travel to the target and back to the sonar (Figure 9.17). The first one is directly from the sonar to the target and back in a straight line. This is the recorded backscatter, from which the slant-range and the ground-range are computed. In certain survey geometries, the acoustic energy can be scattered from the target toward the surface of the sea, and a small portion of it will be scattered back to the sonar ("single surface reflection"). A portion of the incident sonar beam (or its sidelobe) can even scatter toward the surface of the sea, be reflected toward the target, and scattered back to the sea surface and then the sonar ("double surface reflection"). Following these three paths (and therefore three different slant-ranges), the imagery will display two or three images of the same target.

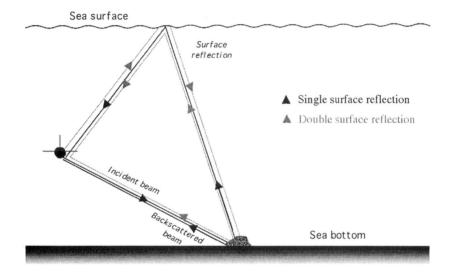

Figure 9.17. Formation of multiples.

Even if it seems somewhat complicated, this effect, also called "multipath effect", is actually observed. This occurs generally in smooth seas, in shallow waters, and when specular reflectors or very reflective targets (e.g. cylindrical navigational aids, pipelines and cylindrical mines or torpedoes) are found on the bottom (Fish and Carr, 1990). Figure 9.18 shows an example of multiple reflection, acquired with SeaMARC II in very smooth sea conditions. The meandering channel which occupies most of the image is partially repeated at far range. It is recognised as a multipath reflection because it repeats the first object at further range and does not fit into the surrounding geological context.

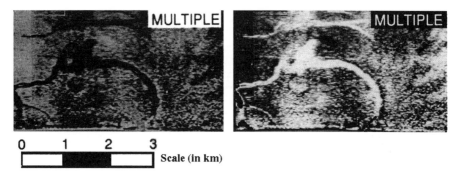

Figure 9.18. Multiple reflections on the seafloor, seen with SeaMARC II on the continental shelf off Peru. The original image (left) has been inverted (right) to show grey levels linearly increasing with the amounts of backscatter. Courtesy R. Hagen, AWI (Germany.)

Multiple reflections are also seen on GLORIA images (LeBas and Mason, 1994). They occur mainly in sedimentary areas where the ocean floor is flat and where bottom reflection is strongly specular. In areas of rough and rocky terrains, such as mid-ocean ridges, the dispersal of the specular component, together with the strong backscatter of the terrain, normally masks multiple reflections. On GLORIA imagery, the first multiple reflections usually manifest themselves as a pair of thin bright lines equidistant on either side of the sonar's track. Pairs of higher multiples may appear at greater ranges. In extreme cases, up to five pairs of multiple reflections can be seen. Algorithms for suppression of multiple reflections can be found in articles by Reed and Hussong (1989) and LeBas and Mason (1994).

9.6.4 Unexpected Features

Interpretation errors may sometimes be related to an incomplete knowledge of the type of local geology. They are, however, generally attributed to unexpected features. Unexpected structures visible on sonar images can include remnants of another type of geological activity, such as old seamounts in abyssal plains. In some cases, they can be attributed to biological activity (schools of fish, whales or dolphins transmitting at the sonar's frequency).

Out-of-range returns are another type of unexpected feature. The returns from the acoustic pings sent by the sonar are received in small time intervals. After transmitting each ping, the sonar receives its echoes and stops after a certain time. But the sonar waves are still propagating out into the water column. They may be backscattered and arrive during the receiving time associated with a later ping. These "late echoes" are attenuated so much that they are usually not recorded by the system. However, very reflective targets just out of range will still manage to be recorded and appear on the imagery. These "out-of-range" returns are very rare, and are generally observed in shallow waters with man-made reflectors such as walls and caissons (Fish and Carr, 1990).

Unexpected structures are likely to be anthropogenic in nature, as no depth environment is any longer devoid of man-made objects. Trawl marks are visible even in the deep abyssal plains. Remnants of other exploration surveys (e.g. drill holes, rig construction remains) can be seen in continental margins. But the coastal environments are the most prone to unexpected anthropogenic features. For example, in the last decade, a private company investigated some parts of Loch Ness (Scotland) with sonar. At regular places on the Loch's floor, they found large circles of stones. These circles could not be explained as geological structures. Inevitably, there started some rumours about "Nessie's nests" (Figure 9.19), huge nests where the Loch Ness mythical beast(s) would live. The truth was, however, much less romantic. Investigations into the past history of the Loch's floor revealed that the last important event was the construction of the Glen Fault Route at the beginning of the century. The large amounts of rocks quarried from the canals linking the lochs together were disposed of by being put on barges and dropped into the lake. These barges opened through the middle, letting the rocks fall to the bottom to form concentric deposits ... .

Figure 9.19. Unexpected features on the seafloor may come from a variety of sources, the most common being anthropogenic activity.

The two images presented at the chapter's beginning (Figure 9.1) are examples of unexpected structures that can be interpreted as anthropogenic, but are purely natural. The "Face on Mars" is a famous image, taken by Viking-1 in the northern latitudes of Mars. In the middle of eroded mesa-like landforms, the optical image shows a huge rock formation, approximately 1.5 km across (Figure 9.20, left). The structure resembles a human head, and has often been misinterpreted as such despite NASA-JPL's explanations. The rock formation is similar to the other ones nearby, and its strange appearance is explained by the illumination conditions. The sun was at ~20° above the horizon, producing important shadows (particularly visible in the south of the structure). These shadows enhance small topographic variations on top of the structure and give the illusion of eyes, nose and mouth. The pixel-sized speckle (black points) originates in errors during the transmission between Mars and the Earth. Images of the

"Face on Mars" taken during later orbits of Viking-1 confirmed the interpretation of the structure as a natural elevated plateau similar to the surrounding mesas.

The other image (Figure 9.20, right) is a TOBI image from the Mid-Atlantic Ridge, north of the Kane Fracture Zone. The image is fully processed for radiometric and geometric distortions. An elongated structure shows up in the middle of the image. It is approximately 500 m long and 200 wide, and looks very similar to the "Face on Mars". In fact, it corresponds to a seamount, itself part of an Axial Volcanic Ridge (see Chapter 5: Mid-Ocean Ridge Environments). The ridge is sub-parallel to the flight path, and ensonified from the left. The high reliefs create important shadows, and small-scale variations in the topography produce small shadows aligned across-track (the"eye", the tip of the "nose", and the "mouth"). A similar structure was seen on Venus during the first Magellan orbit, when the radar imaged at a low angle an elongated seamount with small-scale topography appropriately arranged on the top (the anthropomorphic appearance of the structure also disappeared during the later orbit, when the illumination geometry changed).

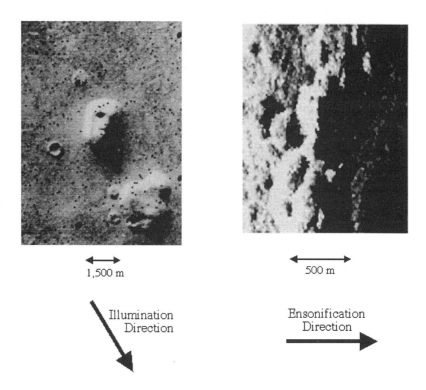

1,500 m   500 m

Illumination   Ensonification
Direction   Direction

Figure 9.20. (Left) the "Face on Mars" (Viking image P-17384-35A72, Copyright NASA-JPL); (right) the "Face on the Seafloor" (TOBI imagery, Copyright SOC).

Another example of anthropogenic activity is the long linear feature found on top of a mud volcano in the Black Sea (Figure 9.19). This image was acquired by the Russian sonar MAK-1 during a survey of mud volcanoes in the Mediterranean Sea (Ivanov et al., 1996) (see section 7.7.1). The telephone cable was apparently laid on the seafloor without any preliminary survey; it crosses perturbed mud flows and the top of the Novorossiysk mud volcano. Telephone cables and pipelines can be seen in any type of seafloor environments (hopefully in a less dangerous position than this one).

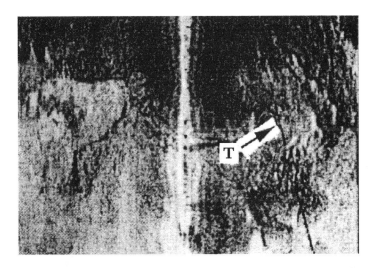

Figure 9.21. Telephone cable (T) on the bottom of the ocean, imaged with the Russian sonar MAK-1. The image is approximately 5 km wide. From Ivanov et al., 1996. Copyright Elsevier Publications.

9.7 CONCLUSION

This chapter is limited in scope and in the number of examples shown. Entire books could be devoted to image anomalies, artefacts, and rare occurrences visible on sidescan sonar imagery. Rather, following the acquisition and processing stages respectively explained in Chapters 2 and 3, all possible sources of errors and misinterpretations have been reviewed, and their causes and effects explained. Some of these problems are quite commonly encountered (e.g. layover), and some are very rare (e.g. interference fringes).

Whenever possible, we have tried to point out possible remedies, either in the operation of the sonar or in the digital processing of the data. Corrections of particular effects specific to certain sonars or operating conditions, and the theoretical background necessary to devise new correction algorithms, are available in the selected references at the end of the chapter, and in the extended General Bibliography at the end of the book.

Hopefully, with the help of the previous chapters, all structures in the sonar images should be interpretable. This interpretation is mostly qualitative; the different structures visible on the seafloor have been shown in their normal context, and can be assigned names and related to their surrounding environment. The progress made in recent years by computer technology enable the interpreter to go further, to quantify the interpretation, and to detect structures or details that were not readily visible before. These techniques now available for sidescan sonar interpretation are presented in detail in the next chapter.

9.8 FURTHER READING

- **About shallow-water surveys:**

Fish, J.P., H.A. Carr; "Sound underwater images: a guide to the generation and interpretation of sidescan sonar data", EG&G Marine Instruments: Cataumet, 189 pp., 1990.
(also available on the Internet : http://www.marine-group.com/acoustic.html)

- **About propagation artefacts:**

LeBas, T.P., D.C. Mason; "Suppression of multiple reflections in GLORIA sidescan sonar imagery", Geophysical Research Letters, vol. 21, no. 7, p. 549-552, 1994

Mason, D.C., T.P. LeBas, I. Sewell, C. Angelikaki; "Deblurring of GLORIA sidescan sonar images", Marine Geophysical Researches, vol. 14, p. 125-136, 1992

Somers, M.L.; "Sonar imaging of the seabed: Techniques, performance, applications", in *Acoustic Signal Processing for Ocean Exploration*, J.M.F. Moura and I.M.G. Lourtie (eds), p. 355-369, Canadian Govt., 1993

- **About interference fringes:**

Huggett, Q.J., A.K. Cooper, M.L. Somers, R.A. Stubbs; "Interference fringes on GLORIA sidescan sonar images from the Bering Sea and their implications", Marine Geophysical Researches, vol. 14, p. 47-63, 1992

10

Computer-Assisted Interpretation

10.1 INTRODUCTION

Interpretation of sonar images, and more generally of remote sensing images, has traditionally been performed visually by trained interpreters. This presents the distinct advantage of using the skill of the interpreter to limits which are yet unattainable by computers. But there also many disadvantages to a purely visual interpretation. First of all, it is subjective: two interpreters with different experience, or different skills, are likely to find slightly different interpretations for some features, depending on their experience of the sonar used or of the environment studied. Visual interpretation is also time-consuming, and a longer amount of time spent on analysis does not ensure a higher objectivity. Structural geologists all know that some morphologic trends will be highlighted, unwillingly and unconsciously, when the time spent on interpretation is too long. The other important disavantage of visual interpretation is that it is qualitative. Objects are outlined, trends and patterns are shown. But their quantitative assessment requires either the interpretation to take place directly on a numeric support (with all the associated problems of small screen size and limited range of scales available), or to scan and quantise the interpretation made on physical supports (paper maps, photographs, etc.).

Computer-assisted interpretation has been made possible by the huge advances made in the last decade by both computer technology (hardware) and its applications (software). It encompasses the fields of numerical processing (geometric and radiometric corrections, as outlined in Chapter 3: Sonar Data Processing), image processing and information management. Computer-assisted interpretation aims at enhancing the visibility of objects, and relations between objects, that were not accessible previously. In some cases, it can also bring information that was invisible to the human eye for physiological reasons (see section 10.4: Texture-Oriented Analysis). Most importantly, computer-assisted information brings an objective and quantitative assessment to help the interpreter. The present chapter does not intend to be a complete guide to all aspects of image processing and computer-assisted interpretation. Many excellent books, referenced in the "Further Reading" section, have already been published on the subject, and the huge number of articles published each month in technical journals shows it to be an ever-increasing and complex field. Rather, this chapter aims at explaining the basics of image processing: what kind of operations can be applied to sonar images, and how they supplement the visual interpretation.

The next section (section 10.2) is devoted to image statistics, and how to enhance the appearance of an image prior to interpretation. As opposed to other domains of remote sensing (e.g. satellite images), and because of the physics involved, the spectral domain is rarely if ever touched upon during sonar surveys. The numerical methods described in this chapter will therefore focus on the spatial domain: the recognition and analysis of structures inside the image (section 10.3), as well as the analysis of textures and patterns inside the morphologic regions of the image (section 10.4). The ultimate goal is of course the identification and interpretation of all distinctive objects in the sonar images. This is often achievable with the use of additional information (section 10.5), such as bathymetry, photographic evidence or other ground-truthing. These results, along with the different levels of interpretation, are then merged into common entities referred to as Geographical Information Systems (G.I.S.). Section 10.6 will explain the common acceptations of G.I.S., show a few examples of marine applications and detail likely developments. Using the techniques described so far, an increasingly important field of study is the classification of seafloor types, for interpretation *per se* and for data compression. This growing area of research is described in section 10.7, along with the state of the art of current international efforts. Artificial intelligence is bound to play as important a role in sonar remote sensing as it now plays in other domains of remote sensing (e.g. radar imaging of the Earth). Accordingly, its potential for computer-assisted interpretation and the establishment of expert systems is outlined in section 10.8, along with the foreseeable developments.

10.2 IMAGE ENHANCEMENT

10.2.1 Image Representation

To be in a form suitable for computer processing, sonar images are digitised both spatially and in amplitude. The spatial digitisation process is called sampling. Because the original sonar images can be too large to visualise and store easily, it may sometimes be useful to reduce the sampling size of the image, therefore decreasing the resolution. For example, a TOBI image originally digitised along 900 columns and 800 lines, with 256 grey levels, will require 703 kilobytes of storage space (Figure 10.1). If the sampling length is decreased by a factor of 3, the image will only require 78 kilobytes (9 times less). By decreasing it again, one can reduce dramatically the space needed for the image. But the original pixel size of 6 metres is similarly degraded to 18 metres and even worse. These effects are shown in Figure 10.1: the original image (top left) has been subsampled by a factor 3 (top right), by a factor 9 (bottom left) and by a factor 12 (bottom right). The first subsampling does not degrade substantially the visual aspect of the image: all the structures around the flat-topped volcano are still perfectly visible. When the sampling is further decreased (Figure 10.1: bottom left image), only the larger structures are visible: the seamount and the fault scarps. The smaller details, such as the hummocky mounds, are not visible. And when the sampling decreases one step further (Figure 10.1: bottom right), only the major faults and the outline of the seamount are recognisable. The degree of discernible detail is strongly dependent on the sampling interval. The latter should therefore be as close as possible to the original footprint of the sonar system.

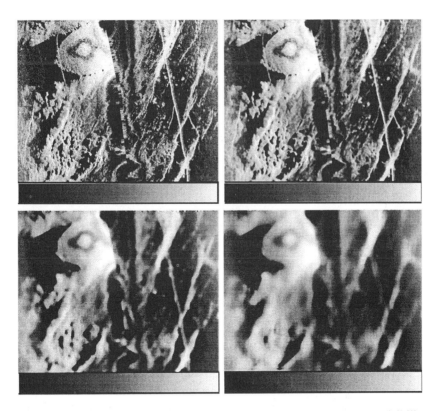

Figure 10.1. Reducing the sampling decreases the resolution and the visibility of small-scale structures. The original TOBI image (top left) has been subsampled by a factor 3 (top right), by a factor 9 (bottom left) and by a factor 12 (bottom right).

The digitisation of amplitudes is referred to as quantisation (see Chapter 3: Sonar Data Processing). The values are resampled to discrete integer values more adequate for computer handling. Most sonar processing packages use 8-bit quantisation, i.e. integer values between 0 and 255 (2^8=256 possible values). This is not always the case, and, before, during or after the processing, the values associated with the pixels may be larger or smaller. The reduction in grey-level dynamics (i.e. the number of bits used for the storage of each value) directly leads to a reduction in size of the whole sonar image. The previous TOBI image, stored with 256 grey levels, will use 703 kilobytes. If the dynamic is reduced to 16 grey levels, it will require half as much storage. And if the number of grey levels goes down to 2 ("binarisation" of the image), it will require ten times less storage! The gains can be important, but the resolution of the image will be concomitantly degraded (Figure 10.2). The reduction to 64 grey levels still produces an image very close visually to the original. With 16 grey levels, all the small details in the contrasted terrains are lost. And the binarisation only produces an outline of the more extreme features in the image: fault scarps facing the sonar beam and shadows.

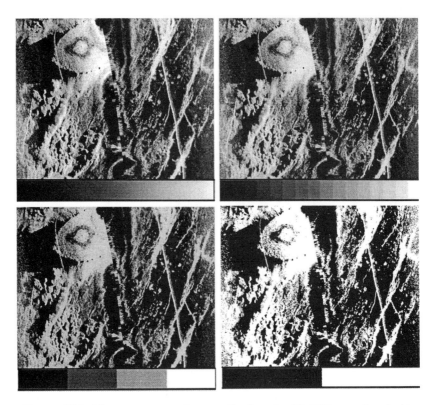

Figure 10.2. The same sonar image, displayed with 256 grey levels (top left), 64 grey levels (top right), 16 grey levels (bottom left) and 2 grey levels (bottom right).

Although simple, these two examples show the effects on image quality. To be susceptible of a rigorous and exhaustive interpretation, the image displayed and stored should be as close to possible to the original image. Modern storage and computing facilities now allow direct analyses of the full-resolution sonar imagery, but the effects of subsampling and dynamics reduction should not be underestimated.

10.2.2 Image Statistics

The information present in the image can be quantified with several measures. We saw some of them (e.g. contrast, mean, median) in Chapter 3 (Sonar Data Processing). The structures visible in sonar images (such as Figure 10.2) are first distinguished by their varying grey levels. The fault scarps, facing the sonar beam, are bright and homogeneous. The seamount's top is grey and mottled. The terraces between the fault scarps are darker and less mottled, and the shadow regions are homogeneously dark. All these regions present different characteristics. An image with Nr rows and Nc columns can be numerically described with the following statistics :

Mean grey level :
$$\bar{g} = \frac{1}{Nr \times Nc} \times \sum_{i=1}^{Nr} \sum_{j=1}^{Nc} g$$

Variance :
$$\sigma_g^2 = \frac{1}{Nr \times Nc} \times \sum_{i=1}^{NG} \sum_{j=1}^{NG} (g - \bar{g})^2$$

Skewness :
$$S_g = \frac{1}{Nr \times Nc} \times \frac{1}{\sigma_g^3} \sum_{i=1}^{NG} \sum_{j=1}^{NG} (g - \bar{g})^3$$

Kurtosis :
$$K_g = \frac{1}{Nr \times Nc} \times \frac{1}{\sigma_g^4} \sum_{i=1}^{NG} \sum_{j=1}^{NG} (g - \bar{g})^4 - 3$$

Energy :
$$E_g = \frac{1}{Nr \times Nc} \times \sum_{i=1}^{Nr} \sum_{j=1}^{Nc} g^2$$

These measures are used to describe the whole image or parts thereof, for operations such as enhancement (see section 10.2.3) or for the delimitation of specific features and regions (sections 10.3 and 10.4). The standard deviation is σ_g, the square root of variance. The skewness measures the shape of the distribution of grey levels. A positive skewness means the distribution is asymmetric with a tail for values greater than the mean grey level. Conversely, a negative skewness means the distribution of grey levels is more biased towards values smaller than the mean (Figure 10.3). The kurtosis measures the relative peakedness or flatness of the distribution, relative to a normal distribution. The value 3 is subtracted from the kurtosis so that $K_g = 0$ for zero-mean Gaussian processes (Pratt, 1978).

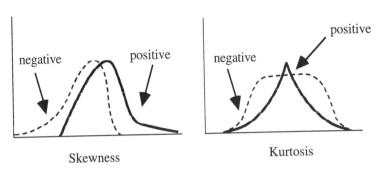

Figure 10.3. Skewness and kurtosis for a distribution of grey levels significantly different from a Gaussian distribution.

First-order statistics quantify the global distribution of grey levels in the image. They can be used to distinguish features or regions which are simple enough to be characterised by a few grey levels only (e.g. well-defined fault scarps always facing the sonar). They are not sufficient for more complex features or separate objects sharing the same grey levels.

10.2.3 Image Enhancement Techniques

10.2.3.1 Histogram Manipulation

The objective of image enhancement techniques is not to increase the quality of the sonar image, but rather to increase the visibility of regions of interest. Common histogram operations were presented in Chapter 3 (Sonar Data Processing): histogram sliding, histogram stretching, and histogram equalisation.

These techniques are, however, limited, because they aim at transforming the original histogram and making it as uniform as possible. Some applications will require higher detail in homogeneous sedimented regions, whereas others will focus on the heterogeneous areas (e.g. volcanic) and not at all on the sedimented regions. The interpreter will therefore need to specify which grey level ranges should be enhanced, and which can remain unchanged. This operation is called "direct histogram specification" and is detailed mathematically in Gonzalez and Wintz (1977). The principal difficulty in applying the histogram specification method to image enhancement lies in being able to construct a meaningful histogram. The first solution is to specify a particular probability density function (i.e. Gaussian, Rayleigh, log-normal, etc.) and then form a "template" histogram by digitising the given function. The second solution is to specify a histogram of arbitrary shape by forming a string of connected straight line segments, which will be displaced interactively (Figure 10.4). Because of its flexibility, the direct specification method (also called "piecewise linear stretching") can often yield better results than histogram equalisation.

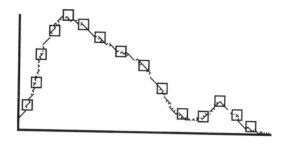

Figure 10.4. The definition of "control points" allows the interpreter to specify which ranges of grey levels need to be enhanced.

Histogram matching applies a predefined set of operations to transform the histogram of an image to resemble the histogram of another image. This process is useful for mosaicking/stencilling (provided the calibrations are similar in both images), and for the detection of changes in the same region (provided the ensonifying geometries are similar).

10.2.3.2 Image Smoothing

Sonar images may present some spurious values due to noise, or more systematic problems during acquisition. These values can be removed by smoothing the image, in the spatial or the frequency domain. Smoothing should be attempted with care, as values corresponding to real data (for example objects at the limit of resolution of the sonar) may as well be removed.

The most simple smoothing method is the neighbourhood averaging. The value of each pixel in the image is replaced by an average of its neighbours. Depending on the size of the neighbourhood, the small-scale details of the image will be more or less filtered. In the case of sonar imagery, a neighbourhood of 3 x 3 pixels is usually sufficient. Mathematically, this corresponds to the convolution of the image by a kernel, or filter. Like the image, the kernel is a matrix of values. The convolution of an image I by a kernel J of dimensions N x P consists in replacing each value I(i,j) of the original image by a new value :

$$I'(i,j) = \sum_{i=1}^{Nr} \sum_{j=1}^{Nc} I(i-k,j-l) \times J(k,l)$$

For example, the image below presents one spurious pixel (white, grey level 255), in the middle of a uniform region (dark grey, grey level 70).

The matrix representing the values of pixels and the kernel matrix respectively are :

70	70	70	and	1/8	1/8	1/8
70	255	70		1/8	0	1/8
70	70	70		1/8	1/8	1/8

The convolution will replace the central pixel by :
$$\frac{1}{8} \times 70 + \frac{1}{8} \times 70 + \frac{1}{8} \times 70 + \frac{1}{8} \times 70 + 0 \times 255 + \frac{1}{8} \times 70 + \frac{1}{8} \times 70 + \frac{1}{8} \times 70 + \frac{1}{8} \times 70$$
i.e. the same grey level of 70 as its neighbours.

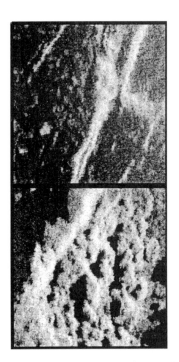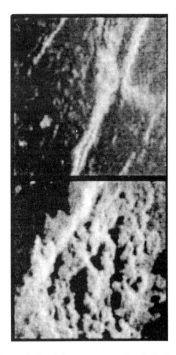

Figure 10.5. Example of smoothing. The original images on the left (top: fault scarps; bottom: hummocky mounds) were imaged with TOBI. Smoothing removes the small-scale noise, but loses some details.

Figure 10.5 presents the application of the same filter to two sample images. The spurious values are indeed removed by the smoothing, but the image is more "blurry". Depending on the image used and the result desired, other filters can be found in the literature (e.g. Gonzalez and Wintz, 1977; Pratt, 1978).

Another way of smoothing the image would be to take advantage of the frequential properties of spurious values. Sharp transitions contribute heavily to the high-frequency content of its Fourier transform. Smoothing can therefore be achieved by removing or attenuating a specified range of high-frequency components in the Fourier space, and computing the inverse Fourier transform of the image. This method is commonly referred to as "low-pass filtering".

10.2.3.3 Image Sharpening

Sharpening techniques are mainly used to enhance the appearance of edges such as morphologic boundaries. Most commonly used is the convolution by a gradient filter. Numerous filters are found in the literature (e.g. Gonzalez and Wintz, 1977; Pratt, 1978), with varying sizes and coefficients.

Figure 10.6. Examples of sharpening.The edges and lithologic boundaries are more visible, but noise is added to the image.

A typical filter would be:
$$
\begin{array}{ccc}
-1 & -2 & -1 \\
+1 & 0 & +1 \\
+1 & +2 & +1
\end{array}
$$

Sharpening enhances the important transitions between grey levels, and, as such, is liable to enhance small-scale noise and spurious values. This is particularly visible in the image shown in Figure 10.6; small blocks in homogeneous regions (terraces and shadows) become more visible, but their shadows are exaggerated.

Since edges and other abrupt changes in grey levels are associated with high spatial frequencies, image sharpening can also be performed in the frequency domain. High-pass filtering of the image's Fourier transform will attenuate the low-frequency components without changing the high-frequency information. Computing the inverse Fourier transform should give a sharpened version of the image.

10.3 CONTOUR-ORIENTED ANALYSIS

The details visible in an image are discernible either as contours (rapid transitions between grey levels) or as textures (the variations of grey levels inside the regions defined by the contours). The contours make up the silhouette of the individual features: spot structures (objects at the limit of resolution of the sonar), line structures (e.g. morphologic boundaries), complete structures (e.g. sedimentation channels, rock outcrops). In this section, some of the most useful techniques for detecting these contours and analysing them statistically will be presented.

10.3.1 Spot Structures

An image spot is a relatively small region whose amplitude differs significantly from its neighbourhood (Figure 10.7). Its size can vary from just one pixel, if the object detected is at the sonar's limit of resolution (e.g. a piece of man-made debris imaged with DSL-120), to a few pixels (e.g. a small seamount imaged by GLORIA). Spot structures can easily be detected numerically. The easiest method compares the value with each pixel with the average value of the other pixels in a small window around it. If the difference is large enough, the pixel is considered as typical spot structure. More elaborate methods are presented in Pratt (1978). One of them consists in smoothing the image with an N x N low-pass filter, and computing the difference between the value of each pixel in the smoothed image and the mean of its four neighbours (right, left, up and down) spaced N pixels away. If the difference is large enough, the pixel is labelled as a spot.

Depending on the geological context and on their size, spot structures may be interpreted differently. They may be spurious values associated to speckle or problems during the acquisition. They may also be genuine structures, such as small boulders, hydrothermal edifices or little seamounts. Statistical analysis of their distribution(s) will help in understanding the local geological processes. For example, small seamounts will be aligned along structural trends and fissures in the Earth's crust (e.g. Lutz and

Gutmann, 1995). And boulders spread along a slope and fanning out will show the extent and importance of a submarine landslide.

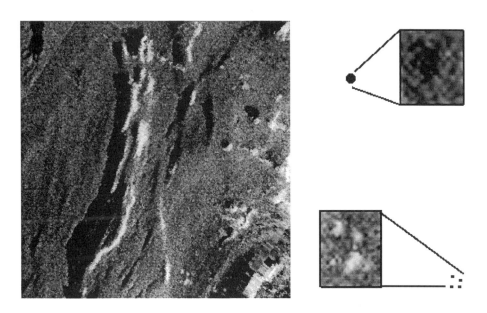

Figure 10.7. Close-up views of spot structures in a TOBI image. The bright features are rock outcrops facing the sonar beam, the dark one is the shadow of an object facing away from the sonar.

10.3.2 Contour Detection

Contours, or line structures, are delimiting regions of rapid changes in grey levels. They are organised arrangements of pixels whose values are differing from the background (Figure 10.8). The simplest structures are connected segments, corresponding for example to fissures on the seafloor. Closed contours (or unclosed if outside the bounds of the image) will delimit morphological regions (e.g. sediment patches, lava flows). Line detection constitutes a branch of image processing in itself, as no method is always 100% satisfactory and the results always depend on the particular application.

The first type of edge detection technique looks at the spectral components of the image. The rapid changes of grey levels occur on short spatial scales, i.e. large spatial frequencies. If these high frequencies can be kept, and the lower frequencies removed or attenuated, the output image should show only the linear structures. In real-world images, the range of spatial frequencies associated with contours will also correspond to small-scale changes inside the regions (local textures), which limits the technique's range.

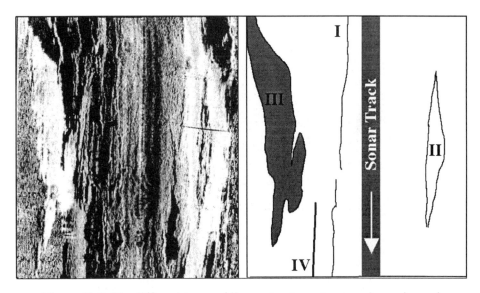

Figure 10.8. The different types of linear structures (contours) are shown in this DSL-120 image: (I) simple line, corresponding to a fissure; (II) closed contour delimiting a lithological region (rock outcrop in the axial wall); (III) unclosed contour, outlining another lithological region; (IV) simple line marking the transition between two regions of differing grey level ranges.

The second type of technique investigates the spatial content of the image, and looks for large changes in pixel values (first and second derivatives). First derivatives are enhanced by gradient filters. The most frequently applied are the Sobel, Prewitt and Kirsch-type filters :

Sobel: -1 -2 -1   Prewitt: -1 -1 -1   Kirsch: -5 -5 -5
            +1 -2 +1   +1 -2 +1   +3   0 +3
            +1 +2 +1   +1 +1 +1   +3 +3 +3

Each of these filters enhances vertical gradients, i.e. the horizontal lines. The enhancement of contours in other directions is accomplished by rotating the matrices around their central element. Figure 10.9 shows filtering of a TOBI image with Kirsch filters respectively oriented NS, NE-SW and NW-SE. Gradient filters enhance particularly well the structures perpendicular to the gradient, but attenuate the other structures. To remedy to this, it is possible to compute the gradients along all directions at the same time and keep only the maximum ones (Blondel et al., 1992). This approach, called adaptive filtering, is shown in Figure 10.10.

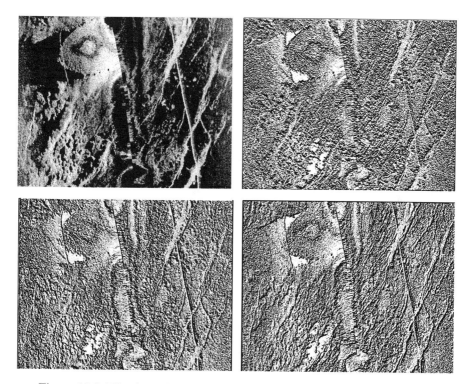

Figure 10.9. Filtering of a sonar image with three Kirsch gradients oriented in different directions.

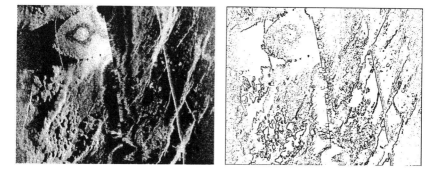

Figure 10.10. Adaptive filtering: gradients are computed in all directions at the same time and only the maximum ones are kept (Blondel et al., 1992).

Changes in second derivatives are enhanced by Laplacian filters. Laplacian filters are used in some recursive filters such as the Deriche filter (Deriche, 1987), but they are highly sensitive to noise (especially point-scale noise). They are therefore often used on pre-smoothed images, at the expense of loss of some information.

$$
3 \times 3 \text{ - Laplacian :} \quad
\begin{matrix}
-1 & -1 & -1 \\
-1 & 8 & -1 \\
-1 & -1 & -1
\end{matrix}
\qquad
5 \times 5 \text{ - Laplacian :} \quad
\begin{matrix}
-1 & -1 & -1 & -1 & -1 \\
-1 & -1 & -1 & -1 & -1 \\
-1 & -1 & 24 & -1 & -1 \\
-1 & -1 & -1 & -1 & -1 \\
-1 & -1 & -1 & -1 & -1
\end{matrix}
$$

Another approach to edge detection is statistical, and looks at the differences between local statistics computed on separate neighbourhoods around each pixel (Rosenfeld and Thurston, 1971). A similar method (Vanderbrug, 1976) uses 14 masks of 5 x 5 pixels to compute the local grey-level variations. These methods are computationally heavy, and the size chosen for the neighbourhoods directly influences the quality of the results. Parallel contours at a few pixels from each other may be detected only once, which is a problem for structures such as ridges and fissures.

Template matching is used when looking for line structures with pre-established shapes. The Hueckel transform consists in moving a circular window of 40 to 100 pixels across the image (Hueckel, 1971; Pratt, 1978). The line segment separating the moving window into two homogeneous regions is considered to be the local segment of the contour. This transform is not sensitive to noise, but requires many computations and gives erroneous results at angles and intersections (Cocquerez, 1984). The Hough transform is used in satellite remote sensing applications to find straight lines (e.g. Wang and Howarth, 1989). The fitting of hypothetical lines to curves in the coordinate space is replaced by fitting to curves in a parametric space (such as the (ρ, θ) space where ρ is the line length and θ its orientation). This method can be applied to more complex structures than straight line segments by increasing the number of parameters (which becomes computationally even more heavy). Hough transforms have been applied to detect seamounts in sonar imagery (Keeton, 1994).

The last approach to edge detection uses the topological properties of contours. It belongs to the discipline of mathematical morphology (Serra, 1982). The features to detect are compared with simple shapes (disks, segments ...), called structuring elements. Mathematical morphology investigates the neighbourhood and connectivity relationships of objects, and uses relatively few operations: erosion, dilation, opening, closing, skeletonisation. Combined, these operations can help detect contour lines (e.g. Meyer, 1978). However, they tend to be highly susceptible to noise, and are not applicable to complex structures. As such, they are not widely used for natural images, unless the data has been previously smoothed and filtered.

10.3.3 Contour Analysis

Once the contour lines have been detected, they can be analysed qualitatively by eye. But it is possible to fully use the computer power to measure objects ("target mensuration", e.g. lengths, sinuosities, directions, ...) and calculate statistics. Different methods have been developed in the last fifteen years, for greatly varied applications:

anthropometry, character recognition, biomedical imagery, seismic analysis, or mapping of roads in aerial images. Very few of them have been applied to sonar images. These techniques can be divided into two families: "blind" methods which use only past information, and heuristic methods which use past information and make assumptions about information to come.

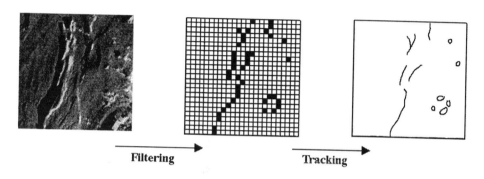

Figure 10.11. Tracking methods aim at transforming the filtered image in an image where the contour lines are properly labelled and quantified.

"Blind" methods assume there is no *a priori* information about the contours. Successive pixels are linked according to logical constraints, without looking for optimal solutions. Three main algorithms have been developed. Generally suited to aerial imagery, the Nevatia-Babu method proceeds in four stages: filtering by the Nevatia-Babu 5 x 5 filters, thinning of contours, search for predecessors and successors of a pixel among its 8 neighbours (3 possibilities only, because of the thinning), and linking of the pixels together. This technique presents a high signal-to-noise ratio, but loses all the small-scale information. Another method, based on the Hueckel operator presented above, investigates the image statistics. The image is swept line by line, and, each time a contour element is encountered, the contour is followed down to its end. In case of multiple successors to a node, the choice is based on local statistics. The third, and most used, of the "blind" methods uses searching patterns. They are pre-established masks that are superposed on the successive pixels in the structure and show where the next successors will be searched (Figure 10.12). Robinson (1977) proposes 72 possible configurations of 3 x 3 masks, which is computationally prohibitive when long structures are present. Cocquerez (1984) uses less searching patterns, of a different type (Fig. 10.12) and adapted to straight features (e.g. roads and electric lines). A less restrictive searching pattern has also been proposed (Blondel et al., 1992), using the information given by gradient filtering about the likely directions of the structures.

Heuristic contour analysis methods directly stem from artificial intelligence applications (see Section 10.8). Particularly complex to use, they call upon graph theory. Each graph is composed of two sets, one with elements called nodes or summits, the other with ordered series of nodes, called arcs. Searching for a contour line is equivalent to searching the optimal route in the graph. Specific algorithms compare the contour being followed to a reference contour. They orient the searches in one direction or another, depending on a cost function quantifying the differences between the two contours. The definitions of the reference contour and the cost function are the main difficulties in producing good results with these methods. If a pipeline can be easily modelled mathematically, a submarine canyon is more complex. Computationally demanding, these methods also generally require high-level programming languages such as LISP or Prolog, not commonly used in the sonar community.

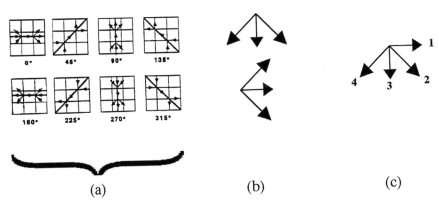

Figure 10.12. Some "blind" methods use specific searching patterns: (a) 8 of the 72 patterns proposed by Robinson (1977); (b) 2 of the patterns proposed by Cocquerez (1984); (3) the generic pattern for searching in the directions given by adaptive filtering (Blondel et al., 1992).

10.4 TEXTURE-ORIENTED ANALYSIS

10.4.1 Texture Definition

Contours enclose areas often characterised by repetitive structures or patterns (Figure 10.13). These repetitions are referred to as textures, and have two main components: statistical (the repetition of a local pattern in a region large in comparison to the pattern's size), and functional or constructional (the decomposition of this pattern in a non-random arrangement of elementary parts with similar dimensions). For example, in Figure 10.13, the leftmost region (sediments) exhibits a "smooth" texture: small variations in grey levels, no discernible spatial organisation. The central region shows a more contrasted image, with a preferential (vertical) direction along which the faults and fissures are oriented. The righmost region corresponds to several hummocky mounds:

the contrast between grey levels is less important, and the sub-circular structures are not aligned along any preferential direction.

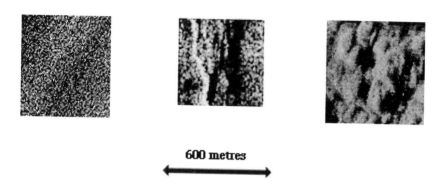

600 metres

Figure 10.13. Examples of different textures discernible in sonar images (from the TOBI image shown in Figure 10.14). From left to right: sediment-covered areas; tectonised area with faults and fissures; volcanic area.

Definitions of texture are quite general and do not lead to simple quantitative measures. Textural measurements can be extracted from the image with various techniques (Table 10.1). Structural methods assume there is an underlying order behind the textures and try to model it mathematically. Statistical methods do not assume any order but measure the variations of local or global textures.

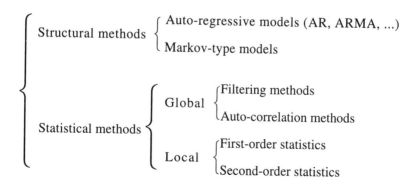

Table 10.1. Texture analysis techniques found in the literature can be divided into two main groups.

10.4.2 Structural Methods

Structural methods assume there is an order underlying the local textures. They endeavour to characterise textures in terms of their primitives (basic elements of textures) and the placement rule governing their arrangement within the image. Auto-regressive (AR) models are quite recent. Working in the frequency domain, they assume that each pixel value is a linear combination of its surrounding values, plus some white noise. The number and values of the parameters in the linear combination define the type of AR filters (Max, 1985). Variants are the ARMA (auto-regressive moving average) and CAR (circular auto-regressive, using a circular window) models. Local textures are defined by the AR models they can be associated with. These methods have proven very useful in fields like robotics (e.g. Wang and Howarth, 1989) but they risk being highly sensitive to the noise always present in sonar images. Other models are more mathematical. Markov random fields represent local interactions between neighbouring pixel values in terms of Gibbs energy functions (e.g. Nguyen and Cohen, 1993). They have been applied to the recognition of basic geological units in sidescan sonar imagery (Jiang et al., 1993) and are increasingly used in other remote sensing applications.

10.4.3 Statistical Methods

First-order statistics quantify the global distribution of grey levels in the image. They are a first approach toward texture quantification (Figure 10.14). Although some geological regions can be coarsely recognised from distinct intervals of grey levels, a natural image cannot be described on the basis of its grey levels alone (e.g. Haralick and Shapiro, 1985; Shokr, 1991; Blondel, 1996). It is more useful to describe the regions by their statistics (i.e. mean, variance, skewness, etc.) (see section 10.2.2).

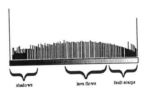

Figure 10.14. In an ideal world, geological features would be discernible on the basis of their grey levels alone. This is true for this particular image, and as far as the analysis does not require fine distinctions between regions. More detailed interpretations need to use first-order statistics or other techniques.

Second-order statistics quantify the spatial relationships of grey levels in the image. Experiments on human vision demonstrated the eye could not always distinguish between textures with different second-order statistics, proving the advantage of computer-based methods. The technique most commonly used in remote sensing studies in general (e.g. glaciology, meteorology, land-cover analyses) uses Grey-Level Co-occurrence Matrices (GLCM). GLCMs address the average spatial relationships between pixels of a small region (Haralick et al., 1973). The textural information is described by a set of several matrices { $P_D(i,j)$ }. Each element $P_D(i,j)$ expresses the relative frequency of occurrence of two points, with respective grey-levels i and j, at a distance $D(d,\theta)$ from one another. If the image is quantified with NG grey levels, the GLCMs will be NG × NG arrays. They are computed on finite windows, of dimension WDSZ × WDSZ. In homogeneous regions, differences between grey levels will be low, and elements close to the diagonal of the GLCMs will therefore have higher values. Less homogeneous regions will yield higher differences between neigbouring grey levels, and resulting GLCMs will therefore have higher values further away from the diagonal. More detailed discussions of the variations of co-occurrence matrices with the images are presented in Haralick et al. (1973) and Shokr (1991). Co-occurrence matrices are not easy to interpret directly, and they are more effectively described by statistical measures called indices. More than 25 textural indices are available from the current literature, and their usefulness to different types of sonar imagery has been effectively demonstrated (Reed and Hussong, 1989; Blondel et al., 1993; Blondel, 1996). Results of their application to TOBI imagery from the Mid-Atlantic Ridge are shown in section 10.7.1 (Image Classification).

10.5 FUSION OF MULTI-SOURCE INFORMATION

10.5.1 Bathymetric Information

Sidescan sonar imagery is greatly enhanced by the presence of bathymetry, if possible at a comparable resolution. Where are the fault scarps and the slopes? Do the local structures observed on the imagery have a bathymetric expression? Are these bright homogeneous terrains bright because they are facing the sonar, because they are extremely rough, or because the radiometric values are saturated? Answers to these questions, and many more, can be found or checked with the precise knowledge of the local topography. Bathymetry can also be incorporated into the processing of sonar imagery, by refining the slant-range corrections and adding terrain corrections to the backscatter laws in use (see Chapter 3: Sonar Data Processing).

The combination of bathymetry and sidescan sonar imagery into three-dimensional images has a definite visual impact (Plate 4, in the colour section). In this example, also shown on the book's front cover, TOBI imagery from the Mid-Atlantic Ridge has been projected on top of multibeam bathymetry. The depths are expressed as colours and their intensities are proportional to the amount of acoustic energy backscattered. Located near the Broken Spur hydrothermal area (29°N), the image in Plate 4 shows a typical portion of a mid-oceanic ridge. The 3-D view enhances the appearance of the seamount on the left, cross-cut by faults and with highly reflective lava flows at its base. The axial valley in the middle of the image is seen as elongated and flat, bordered on the far left by the axial walls (less visible on the right). On the right, the alignment of seamounts along the axis of the ridge is perfectly visible. This shows the scientific addition of three-dimensional visualisation, especially when compared with the corresponding bi-dimensional images of bathymetry and sidescan sonar imagery (Plate 5, in the colour section). Another interest of merging the two types of data is the possibility of noticing (and correcting) mismatches between coregistered bathymetry and sidescan sonar imagery. Alone or merged into "fly-through" movies, these 3-D presentations help in visualising the 3-D geology. One of the authors (Ph.B.) participated in experiments made at the University of Washington where DSL-120 imagery and bathymetry were merged, and where the geologist could interact through virtual reality and "fly" everywhere in the dataset. An immediate advantage was the visualisation of small-scale features like the hydrothermal vents, barely discernible on the sonar imagery alone. They could then be readily compared with their visual aspect during submersible dives, using the same view angles and trajectories.

Additional sources	Scale	Resulting information
Geological maps from previous surveys	Large-scale	Geological context at the time of these surveys
Magnetics	Medium-scale	Age of the seafloor and localised heterogeneities
Gravimetry	Medium-scale	Density of the seafloor
Seismics	Medium-scale	Subsurface structures
Water column data	Small-scale	Presence of active processes such as hydrothermalism or turbidity currents
Dredges	Small-scale	Geochemistry and surface cover (e.g. serpentinisation, hydrothermal deposits)
Cores	Fine-scale	In-depth geochemistry at specific points
Profiler data	Fine-scale	Surface sedimentation and very shallow subsurface structures
Optical data	Fine-scale	Ground-truthing

Table 10.2. Additional sources of information, other than bathymetry, can help to better constrain the interpretation.

10.5.2 Other Information - Ground-Truthing

Other types of information can be useful to constrain the interpretation (Table 10.2). Previous surveys, at a larger scale, may have produced geological maps, so that one knows for example that the terrain being interpreted is in a sedimentary context and on a continental slope. Additional geophysical data include magnetism (from aerial, surface, or water column data), gravity or seismics, which provide some insights on the larger-scale structures. Vertical probes or deep-towed vehicles provide some physical/chemical information about the processes at play in the water column: heavy currents, large presence of hydrothermally originated particles in suspension in the water, high temperatures in specific spots, etc.

Dredges take samples along a linear portion of the seafloor, and provide some information about the surface (e.g. presence of serpentinites or hydrothermal deposits). The positions of the samples are not precise. Vertical cores give point information which is more precise and specific. More systematic information about the sedimentary cover and the eventual presence of shallow subsurface structures is available from analysis of the profiler data usually recorded simultaneously with sidescan sonar imagery (see Chapter 6: Abyssal Plains and Basins). Finally, the most sought-after source of additional information is optical ground-truthing, whether it be from manned submersibles (e.g. *Nautile*), deep-towed platforms (e.g. Argo-II) or autonomous vehicles (e.g. JASON).

10.6 GEOGRAPHIC INFORMATION SYSTEMS

10.6.1 GIS Definition

The concept of Geographic Information Systems dates back to the 1970s, but it only became popular when the technological means to back it up (hardware and software) started to be accessible to all. In the last part of the 1980s, Geographic Information Systems (GIS) turned out to be an unavoidable concept, sometimes used and misused for marketing purposes more than scientific ones. What in fact are Geographic Information Systems ? Their first acceptation is that of multi-relational databases, "*a powerful set of tools for collecting, storing, retrieving at will, transforming and displaying spatial data from the real world*" (Burrough (1986), cited in Maguire et al., 1991). Their second acceptation focuses more on the final products from a GIS: decision-making and problem-solving, in a spatially referenced environment.

The technological setting of GIS is changing rapidly. Current GIS software is now mostly used on Unix-based platforms, but hardware ranges all the way from laptop computers to mainframes. The number of GIS packages and applications is increasing exponentially, as can be seen in the technical literature (journals such as *GISWorld* or the *International Journal of Geographic Information Systems*). The GIS market is one of the fastest growing computer application markets, with annual markets of several billion dollars and expected growth rates of around 20% per year. And Geographic Information Systems have now permeated all domains, scientific and non-scientific.

10.6.2 GIS Composition

Geographic Information Systems are made of spatially referenced features. The geographical element, also called locational or spatial element, provides the position (XYZ coordinates, such as latitude, longitude and depth below sea level). It is associated with a description of the feature, usually called attribute, or, sometimes, statistical or non-locational data element. For example, the point location of a core sampling is a geographical element used to provide a reference for a geophysical attribute describing the type of seafloor. Geographic Information Systems use indiscriminately point, raster or vector data. Examples of point data are locations of samples and the associated results (e.g. type of seafloor, biological activity recorded, current measured). Examples of vector data are the vehicle's navigation, bathymetric contours, pipeline routes, etc. Vectors with identical start and end define areas, and are often referred to as polygons. Examples of polygons are political boundaries such as the Exclusive Economic Zone, or zones of homogeneous surface geology. Examples of raster data are images of sonar backscatter such as those presented in this book, bathymetric maps, etc. These different types of representation are usually stacked concurrently and called layers. In a typical GIS application, the first layer would be the sonar imagery (raster), the second layer the bathymetric contours (vector), a third layer the location of cores (point) and a fourth layer the geological interpretation (vectors and polygons).

These styles of representation have different advantages, depending on the application. Vector data are feature-based, while rasters are cell-based: vectors therefore represent shapes more accurately (Figure 10.15). But they sharpen the boundaries, whereas raster images are able to represent gradual transitions between regions. The other main difference is the computer cost of these representations: vector data require less storage space but are more demanding computationally and slower to process than raster data.

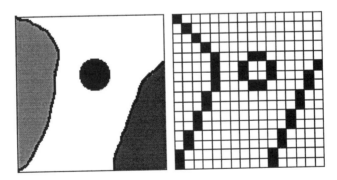

Figure 10.15. Stylised comparison of vector (left) and raster (right) representations of the same structures. Vectors represent shapes more accurately, but rasters can represent gradual transitions between regions.

10.6.3 GIS Operation

How do Geographic Information Systems work? Rhind (1990) and Maguire et al. (1991) distinguished six types of problems that could be solved with GIS (Table 10.3). The location question consists in querying the database to determine the types of structures present at a certain place (e.g. what is the structure at this latitude and longitude?). Conversely, the condition question consists in finding the location of specific features (e.g. where are the fault scarps less than 1 km away from the pipeline route?). The trend question consists in monitoring how things have changed over time (e.g. what is the change in sediment distribution in this area of the harbour?). The other problems are more complex and require spatial analysis as well. The routing problem involves computation of the best route (safest, shortest, flattest ...) between several points (e.g. what is the safest and shortest route to lay a cable?). The pattern problem involves the description and comparison of the distribution of specific structures, in order to understand the underlying processes accounting for their distribution (e.g. is there some common geological setting to all hydrothermal deposits?). The modelling problem consists in simulating the answers to different models (e.g. which areas will be affected by a submarine landslide?).

1	Location	What is at ... ?
2	Condition	Where is it ... ?
3	Trend	What has changed ... ?
4	Routing	Which is the best way ... ?
5	Pattern	What is the pattern ... ?
6	Modelling	What if ... ?

Table 10.3. Basic problems that can be investigated using Geographic Information Systems (after Rhind, 1990, and Maguire et al., 1991).

10.6.4 Examples of GIS Applications

10.6.4.1 Crustal-Scale Fissuring in the East Pacific

The best way to present the concepts outlined in the previous paragraphs is through the presentation of Geographic Information Systems in marine applications. In late 1989, scientists from the University of California, Santa Barbara, and from the Woods Hole Oceanographic Institution, conducted a cruise on a small portion of the East Pacific Rise (Wright et al., 1995). This cruise was investigating small-scale fissuring and its relationship to magmatic and hydrothermal processes. The survey used the *Argo-I* seafloor imaging system. Acoustic and optical data were gathered along multiple, closely spaced (10-50 m) tracklines along ~ 80 km of the East Pacific Rise (EPR) crest. The 100-kHz sonar provided 90% coverage with an 800-m wide swath of terrain centred on the EPR axis. The images provided by the *Argo I* downward- and forward-looking video cameras, with a swath of approximately 13 m, covered more than 9 km^2 of the axial zone. The density of these datasets enabled to determine the distribution of hydrothermal vents and deposits along the axial zone relative to fine-scale volcanic and tectonic features and to provide observations required for evaluating the area's potential for future scientific drilling.

The video footage was double-checked against handwritten cruise logs to ensure that all features visible were properly referenced and located in the GIS database. Fissure widths were calculated on each and recorded as attributes along the vectors representing the fissures. To account for the 5-m uncertainty of the long-baseline navigation, the population of fissures was filtered and only one fissure kept when two or more of them were within 10 m of each other in an east-west direction. Fissures large enough to be located with sidescan sonar (widths > 5 m) were added to the GIS database and represented as lines. Fissures appearing as thin black strips against a dark grey background indicate little or no vertical offset across-track (see Chapter 5: Mid-Ocean Ridge Environments). Thin black strips that are fairly sinuous and show thin or muted acoustic returns on one or both edges were interpreted as large lava channels and therefore not recorded. Finally, relative ages of lava flows were computed from the video footage, on the basis of their respective morphologies and sediment accumulation patterns. The ages were digitised into the GIS as nodes, representing the starting and ending of along-track video observations of flows of a particular age. These lines were then dilated to an arbitrary width of 300 m (the swath width of the *Argo I* sonar).

Spatial relationships were computed with the GIS, enabling the display of distance, length, width, area, abundance, etc. for all items in the database. From these analyses, it appears that cracking in this region is driven by the injection and propagation of dikes rather than by far-field plate stresses. Covariations in fissure density and axial lava ages reveal the tendency of the crust to form more cracks with time rather than widen already-existing cracks. Active hydrothermal vents are most abundant along portions of the axis that have experienced recent dike intrusion. These results are now used to detail the models for the evolution of the segments of the fast-spreading East-Pacific Rise. Without the combination of sonar and optical imagery in a Geographic Information System, it would not have been possible to assess statistically their significance, and how representative or widespread these conclusions were.

10.6.4.2 *Distribution of Hydrothermal Deposits in the North Atlantic*

The applications of Geographic Information Systems in marine environments are of course not restricted to imagery alone. In recent years, the authors participated in a multinational research project (MARFLUX/ATJ) funded by the European Community. This programme aims at investigating volcanic, tectonic and hydrothermal processes along 500 km of the Mid-Atlantic Ridge south of the Azores, close to European shores and partly in Portuguese waters. A Geographic Information System was established in collaboration with the Irish remote sensing company ERA-MAPTEC Ltd. (Critchley et al., 1994). It was based on a compilation from the geophysical data gathered during previous surveys, and updated after a cruise in 1994, during which the first version of the GIS was used as a shipborne decision support system. The data used originated from various sources: acoustics, magnetics, gravity, seafloor and water column samples, biological observations, etc.

The base for the GIS was the compilation of all multibeam bathymetry ever acquired between 34°W-24°W and 35°N-42°N. Bathymetric measurements were added as raster imagery to the GIS database, with a vertical precision of 10-20 metres, and a horizontal resolution of about 100 metres. Centred on the axis of Mid-Atlantic Ridge, the sidescan sonar imagery came from two sources: older GLORIA data (see Chapter 2: Sonar Data Acquisition), and recent TOBI data from the 1994 cruise. Aeromagnetic data (maps, profiles, isochrons) from previous surveys was compiled with sea-surface magnetic anomalies from towed magnetometers, and 3-D measurements from a three-component magnetometer fitted on the deep-tow TOBI vehicle. Attributed to thermal remanent magnetism in the basaltic oceanic crust, magnetic anomalies are used for absolute dating of the seafloor. They were supplemented with gravity measurements from the sea surface and from satellites. Along the cruise, deep-towed platforms and dynamic hydrocasts also provided measurements, respectively along-track and vertically, of chemical concentrations and water column characteristics (nephelometry, salinity, etc.). These latter measurements enabled the detection of 8 new hydrothermal sites in this portion of the Mid-Atlantic Ridge.

Statistical analyses were performed for all layers in the GIS, separately and together. The northern segments are more affected by volcanism, and the southern segments by tectonics. This is attributed to the declining influence of the Azores hot-spot with distance. Cross-correlation of the different processes with each other, with bathymetry, and with the distribution of known hydrothermal sites was further investigated (Blondel, 1996). Bathymetric highs are correlated to volcanism, which is consistent with the construction of the ridge by magmatic processes. And tectonic and volcanic processes are inversely correlated in most of the segments. Using these analyses, it is speculated that hydrothermal activity is spatially correlated with the local tectonics and not necessarily with neo-volcanism. This complements the current models, according to which the increase in activity associated with plate reorganisation may be related to tectonism which creates permeability by rigid plate deformation, and the observations made *in situ* that venting focuses in areas of important cross-cutting tectonics.

10.7 IMAGE CLASSIFICATION - IMAGE COMPRESSION

10.7.1 Image Classification

The volume of data collected by sidescan sonars increases with their resolution. To ensure efficient utilisation of all data, new techniques must be devised for extracting various information and supplementing the interpreter with reliable quantitative material. The presentation of data sets as mere images is not enough, and these new techniques must support with interpretation with quantitative arguments going as far as possible beyond human capabilities. Image classification techniques aim at recognising the different morphological units present in an image, and displaying them as an interpreted map. The fields of application include hydrography, engineering, environmental surveys, geology and mine counter-measures.

Image classification is an open subject of research, and there is no definitely proven and fault-proof technique. Quantitative measures, or indices, are attached to each point in the image (e.g. local contrast, roughness of the local texture). The process of classification relates the indices to the corresponding geological interpretations, and organises in "classes" the multi-dimensional space of the indices. There is no theory of image classification *per se*, as emphasised in the exhaustive review written by Haralick and Shapiro (1985). Classification techniques are basically *ad hoc* and judged on their performance.

Unsupervised classification relinquishes all control on the final classification to the algorithm. The n indices measured at each point will be submitted to clustering algorithms to determine the natural groupings of points in the n-dimensional space. Classic clustering algorithms start from an arbitrary partition of the n-dimensional space into groups, and rearrange these groups to minimise or maximise certain parameters. Known algorithms include K-Means (e.g. Reed and Hussong, 1989) and simulated annealing (e.g. Selim and Alsultan, 1991; Nguyen and Cohen, 1993). Iterative partition rearrangement schemes such as ISODATA have to go through the image many times, which can take excessive computation time. The accuracy of some unsupervised classification schemes depends on the number of regions allowed at the beginning. If the actual number of independent morphologic units is bigger than the starting number of units, the algorithms will produce erroneous results.

Supervised classification techniques use the experience of the interpreter, and/or the reference with calibrated portions of the seafloor, to guide the recognition of units. It is important that all classes are represented, and that the samples used to "train" the algorithms do not represent mixed units. All other sets of n indices will be associated with the classes to which they are most similar. Several classification algorithms are available: minimum-distance classifier, Bayes maximum-likelihood (Duda and Hart, 1973). Measurement-space guided (MSG) clustering is a technique particularly adapted to natural images (Gonzalez and Wintz, 1977; Haralick and Shapiro, 1985), especially in complex and noisy environments such as in sonar images of mid-oceanic ridges. The MSG technique partitions the measurement space and each pixel in the image can be assigned the label of the cell in the partition ("labelling" process). The regions in the image are then defined as the connected components of the pixels having the same label.

Plate 6 (see the colour section) shows an example of a simple classification using first-order statistics. The image was acquired with TOBI and shows a Non-Transform Discontinuity on the Mid-Atlantic Ridge (see Chapter 5). Several swaths are mosaicked together. This portion of the seafloor is dominated by large-scale tectonic processes, as examplified by the large fault scarps everywhere in the image. Small-scale volcanism manifests itself as hummocky mounds at the base of the scarps. Only two flat-topped seamounts are visible near the middle of the image. The maximum grey level, the mean grey level and the standard deviation were computed for each neighbourhood of 3x3 pixels. The supervised classification algorithm (K-Means) used the ground-truthing available for specific regions. The classified units are shown in colours, dark blue corresponding to pelagic sediments, green to chaotic sediments, red to talus scarps, yellow to neo-volcanics, and black to areas with shadows or no data.

Another example of classification is shown in Plate 7 (colour section), this time using second-order statistics. The TOBI image is of a mid-oceanic ridge discontinuity, and shows one high-resolution swath (6 km wide). Textural indices are computed with Grey-Level Co-occurrence Matrices. Entropy measures the roughness of local acoustic textures, and local homogeneity which measures the degree of organisation of the texture. These two indices were demonstrated to be sufficient to describe the main geologic units (Blondel et al., 1993; Blondel, 1996). Measurement-space guided clustering was found to be the most suited to this type of study, in particular because of its tolerance to localised heterogeneities inside homogeneously textured regions. The classified map shows neo-volcanic provinces in red, tectonic areas in blue, and sedimented areas in green. Subdivisions of these units are also possible. Checked with intensive ground-truthing in the Pacific and Atlantic Oceans, this type of classification presents high accuracy ratios (64% to 100%).

Image classification is very important for mapping and interpretation purposes. It can also be useful for changeability studies, for example investigating variations of the environment before and after off-shore drilling, or looking at the redeposition of sediments after a storm. By representing the image with a finite number of morphological units, image classification can also be used for image compression.

10.7.2 Image Compression

Each sidescan survey produces large amounts of data, including at least imagery, navigation and attitude. As outlined in Tables 2.1 and 2.2, the most recent sonars daily return close to 1 gigabyte. This data needs to be transmitted from the sonar to the recording instrument, through a cable with a finite bandwidth, and thus a finite transmission rate. The data needs to be stored and transferred between computers. Even high-density magnetic tapes and optical disks are only limited to a couple of gigabytes. Storage space is more accessible, but most users cannot afford more than 10 gigabytes. Finally, the data needs to be processed and interpreted. A final map of TOBI imagery, processed and produced at a scale of 1:100,000 typically corresponds to 330 megabytes, i.e. 3.3×10^8 individual points.

Basic image compression can be attained by restricting the number of bits on which each pixel is stored (i.e. reducing the radiometric range), by subsampling or averaging the data (i.e. decreasing its original resolution), or by coding recognisable features (and thus maintaining the original quality). Advanced compression algorithms look at the repetition of values or series of values in the image and code them accordingly: Run-Length Encoding, adaptive Lempel-Ziv compression (used in Unix systems), cosine-transform (used for videophone images), etc.

These methods do not take into account the two-dimensional aspect of sonar imagery. Using image classification, objects and regions can be coded more efficiently. For example, fishery applications do not require precise knowledge of the seafloor and a simple definition such as rock/gravel/sand is largely enough. The compressed and coded data can also be used to lighten the burden of a human interpreter looking for specific targets on the seafloor.

10.8 ARTIFICIAL INTELLIGENCE - EXPERT SYSTEMS

A new class of computer has emerged in the last years, with dedicated hardware or software which can process equally well numbers and symbols. The best-known of these systems are the expert systems, coming into use in industry. These systems are derived from the rapidly advancing research in artificial intelligence (AI). Although they are widely referenced, these systems are not always perfectly understood. This short section therefore aims at defining artificial intelligence and how it can be applied to the interpretation of sonar imagery.

Artificial intelligence is often assimilated to the reproduction of anthropoid reasoning with a computer. Human intelligence involves both the ability to solve a specific class of problems, and the latent ability to discover solutions for a new class of problems. These two types of capability are still quite distinct in machines. AI-based systems can be divided into three types: routine-oriented, rule-based, and knowledge-based.

Routine-oriented systems get standard data, and perform simple, repetitive tasks. Neural networks are a good example (Figure 10.16). They consist in nodes, called neurons, and weighted paths connecting these neurons. Each node has an activation level, function of the activation of the nodes directly connected to it. The number of nodes, the number of layers, and the way they are connected define the types of neural networks. Back-propagation or BP networks (e.g. Stewart et al., 1994) and Multi-Layer Perceptrons (e.g. Michalopolou et al., 1995) are used for certain seafloor characterisation studies. These neural networks are calibrated ("trained") with samples of known seafloor types. For each new environment, or each change in the sonar configuration, these neural networks will need to be retrained. They do not have the capability to adapt automatically to new conditions and find new decision rules.

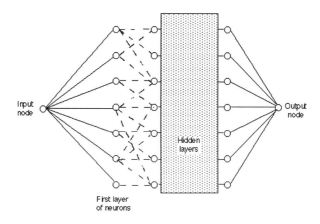

Figure 10.16. Schematic representation of a neural network.

Systems using the rule-based approach focus on the definition of appropriate rules, and their execution in limited specific situations. These systems are commonly referred to as "expert systems", since they codify the rules used by a human expert for a particular task. There are some limitations, as it is difficult to assess the effect on the whole system of adding or changing a few rules. Expert systems are nonetheless used with success in several other domains of industry (e.g. for assessment of hydrocarbon reservoir capacities).

These systems do not have the ability to really learn independently, or to create new concepts by abstracting from experience or by rearranging pieces of knowledge in novel ways. Knowledge-based systems are concerned with more general problems of the organisation of knowledge, and try to enhance the cooperation between various sources of knowledge in a synergistic way. This type of system is often referred to as "knowledge amplifiers".

Applications of artificial intelligence to sonar-related problems have been very few so far. To the best of our knowledge, there are no commercial applications available yet. Neural networks have been used with variable success for acoustic characterisation of the seafloor. Mitchell and Spencer (1995) demonstrated the potential of artificial intelligence for the automatic finding of routes for submarine cables or pipelines in complex terrains. Prospective applications of AI include terrain recognition and terrain avoidance for deep-sea vehicles, and adaptive processing of incoming sonar data. Although still small, the place of AI in marine science and technology is slowly growing.

10.9 SUMMARY

This chapter is intentionally not as detailed as it could be, because it only aims at presenting the basic techniques of computer-assisted interpretation. These techniques have been compiled in a logical order to present how they can intervene at the different stages of interpretation, how they work, and what kind of results they can bring to help the interpreter in its task. We saw how the visual appearance of sonar imagery could be enhanced and small defects corrected (section 10.2) and how to recognise numerically the structures present in the image (section 10.3 and 10.4). Joint analysis of sonar data with other types of information was presented in sections 10.5 and 10.6. This led to the ever-current research conducted in the fields of image classification, image compression (section 10.7), and the envisaged development of expert systems (section 10.8).

One may wonder why computer-assisted interpretation was not used in the previous chapters, or only on a few occasions. The main reason is that we wanted to show the thought processes at play behind the interpretation of images. The second reason is that not all studies require the use of numerical techniques, because of time constraints or budget constraints. But it is indisputable that advanced interpretation of sonar images can only benefit from the techniques outlined in this chapter. The different disciplines constitutive of computer-assisted interpretation are growing rapidly and the next decade should see their rapid development in industrial and academic applications alike.

10.10 FURTHER READING

- **About image processing:**

Blondel, Ph.; "Segmentation of the Mid-Atlantic Ridge south of the Azores, based on acoustic classification of TOBI data", in MacLeod, C.J., P. Tyler and C.L. Walker (eds), "Tectonic, Magmatic and Biological Segmentation of Mid-Ocean Ridges", Geological Society Special Publication, no. 118, p. 17-28, 1996

Blondel, Ph., C. Sotin, Ph. Masson; "Adaptive filtering and structure tracking for statistical analysis of geological features in radar images", Computers and Geosciences, vol. 18, no. 9, p. 1169-1184, 1992

Haralick, R.M.; "Statistical and structural approaches to texture", Proceedings of the IEEE, vol. 67, no. 5, 1979

Haralick, R.M., L.G. Shapiro; "Image segmentation techniques", Comp. Vision, Graphics and Image Processing, vol. 29, p. 100-132, 1985

Max, J. ; "Méthodes et techniques de traitement du signal et applications aux mesures physiques", vol. I, Masson, Paris, 354 pp., 1985

Pratt, W.K.; "Digital image processing", Wiley-Interscience, New York, 750 pp., 1978

Serra, J.; "Mathematical morphology", Academic Press, New York, 610 pp., 1982

- **About Geographical Information Systems:**

Critchley, M.F., D.W. Coller, C.R. German, Ph. Blondel, C. Flewellen, L. Parson, I. Rouse, D. Teare, H. Bougault, D. Needham, M. Miranda; "Integration of deep tow sidescan sonar imagery, bathymetry and other data along the Mid-Atlantic Ridge", EOS Trans. AGU, p. 579, Vol. 75, no. 44, Nov. 1994

Maguire, D.J., M.F. Goodchild, D.W. Rhind; "Geographical Information Systems - Volume 1: Principles", Longman - Wiley and Sons, 649 pp., Harlow, UK, 1991

Legg, C.; "Remote sensing and Geographic Information Systems - Geological mapping, mineral exploration and mining", Praxis-Wiley, 166 pp., Chichester, UK, 1995

Wright, D.J., R.M. Haymon, D.J. Fornari; "Crustal fissuring and its relationship to magmatic and hydrothermal processes on the East Pacific Rise crest (9°12' to 54'N), Journal of Geophysical Research, vol. 100, no. B4, p. 6097-6120, 1995

- **About artificial intelligence:**

Simmons, A.B., S.G. Chappell; "Artificial intelligence - Definition and practice", IEEE Journal of Oceanic Engineering, vol. 13, no. 2, p. 14-42, 1988

11

Conclusion

Sonar, or the use of sound as a means of communication, navigation and perception, is not unique to human activity. Rather, it is from the animal world that humans have borrowed the concept of sonar. Whales use sound to communicate with oneanother. In fact they exploit the sound velocity variation in the upper water column as a waveguide to transmit their cries across hundreds of kilometres. Dolphins and many smaller aquatic mammals use high-frequency sonar to investigate the properties of objects and prey in the water. By chirping their frequencies, dolphins are able to get multi-spectral data about the shape, composition and position of these mid-water targets. Our attempts to mimic the power of the natural world have led to the development of echo-sounders, of sidescan sonars and, recently, of multibeam mapping systems. However, it is the development in recent years of two critical technological aspects that have opened up the world of the deep-sea floor: high-resolution sidescan sonar and computer-aided image processing and interpretation. These technologies are permitting us an unprecedented view of our planet's surface: the "inner space" of our own environment that has hitherto remained a mystery.

Remarkably, in the past ten years, large areas of the deep-sea floor have been imaged by a number of different sidescan sonar systems. Many different geological environments have been explored, shedding light on deep-marine processes for the first time. The importance of understanding our marine environment is paramount, if we are to predict anthropogenic effects on the planet, to exploit the resources of the deep-sea floor, and to generally explore and discover the hidden two-thirds of the Earth's surface. Yet despite the recent advances in sonar imagery, there have not been any concise or comprehensive reference works about this innovative form of remote sensing. The few text-books available are generally dated, and vary between highly technical manuals that concentrate on acoustic theory and highly focused geological works.

In this book, we have intended to provide a general reference covering the entire range of marine seafloor environments. This work is aimed at a range of users: from the professional geologist, environmental scientist and marine engineer, to the acoustic technician and computer scientist. The intention is not to provide an exhaustive atlas of every marine geological feature, but to illustrate a range of features imaged by different systems and educate the reader in the art and science of sidescan sonar interpretation. We have presented three main areas: (1) data acquisition and processing; (2) the geological environments; and (3) realising the limits of the data and enhancing the interpretation with computer and numerical techniques.

The data acquisition sections detail the latest state of sidescan sonar instrumentation, giving technical specifications, types of usage and indicating suitability for a range of survey options. With so many systems available in such a wide range of operational conditions and requirements, the choice of a particular system is the key to successful investigations. We have attempted to illustrate a range of marine environments that span the full depth range found on Earth: from the deep-ocean trenches of more than 10 km depth, through the abyssal plains and mid-ocean ridges to the shoaling continental slopes and near-shore continental shelves. For the first time, some of the best and most recent examples of sidescan sonar imagery from these environments have been published together within a single text. For this, we are indebted to the many scientists and technicians for their cooperation in making these stunning images available. Finally, we have illustrated the most recent developments in sidescan sonar processing and automated interpretation. The rapid advances in computer technology mean that many of these processing routines will be available to an ever-widening community. However, the basis behind the processes remains the same, regardless of the configuration of the computer hardware. We hope that by describing the state of the art of sonar processing, many more users will enjoy the advantages of automated processing and interpretation.

Despite our best attempts, this book remains limited in its scope. There are many more images of the deep-ocean floor that we would have liked to publish but did not have the space for; new sonar systems are being developed all the time; and new ways in which to present the data are being invented. It was not always possible, nor was it the purpose of this book, to examine in detail the geological significance of the varied structures found on the seafloor. To help compensate for this, we have included a small bibliography (Further Reading) at the end of each chapter, with background articles or books. For readers wishing to go further, we have supplemented the "Further Reading" sections with an extended bibliography at the end of this book. Similarly, we would have liked to describe the many other, varied, ways in which sonar is used in the marine environment: acoustic navigation, data telemetry, multi-frequency echosounders, mid-water imagery, sector-scanning sonars and acoustic determination of geotechnical properties of the seafloor. The field of sonar is expanding at an exceptional rate, driven onwards by the need, scientifically, environmentally and commercially, to understand the marine environment.

In conclusion, this book is not a totally comprehensive and exhaustive text on sonar. However, it is the first of its kind since the advent of sidescan sonar remote sensing (or: the study of the sea floor with sidescan sonar), and we sincerely believe that it will be as useful to experts as a source of reference, or to newcomers as an introduction to sonar.

General Bibliography

The following references are arranged by chapters. They do not pretend to exhaustivity, as there are several thousands of articles dealing with the different subjects evoked in this book. But they are what we think are the most important articles or books for the further study of the themes addressed in each chapter.

Chapter 2: Sonar Data Acquisition

Blackinton, J.G.; "Bathymetric resolution, precision and accuracy considerations for swath bathymetry mapping sonar systems", IEEE Proc., p. 550-556, 1991

Blondel, Ph.; "Traitement et Interprétation des données radar: applications à l'étude de la surface de Vénus", (*Processing and interpretation of radar data: Applications to the study of the surface of Venus*), Ph.D. Thesis, 327 pp., Univ. Paris-VII Jussieu, France, 1992

Blondel, Ph., L.M. Parson; "Sonar Processing in the U.K.: A short review of existing potential and new developments for the BRIDGE Community", 27 pp., BRIDGE Position Paper no. 1, Natural Environment Research Council, UK, 1994

Brekhovskikh, L.M., Yu. P. Lysanov; "Fundamentals of Ocean Acoustics", 2nd edition, Berlin: Springer-Verlag, 270 pp., 1991

Bruce, M.P.; "A processing requirement and resolution capability comparison of sidescan and synthetic-aperture sonars, IEEE J. Oceanic Engineering, vol. 17, no. 1, p.106-117, 1992

Chatillon, J., M.E. Bouhier, M.E. Zakharia; "Synthetic Aperture Sonar for seabed imaging: relative merits of narrow-band and wide-band approaches", IEEE J. Oceanic Engineering, vol. 17, no. 1, p.95-105, 1992

Denbigh, P.N.; "Swath bathymetry: Principles of operation and an analysis of errors", IEEE J. Oceanic Eng., vol. 14, no. 4, p. 289-298, 1989

Fleming B.W.; "Sidescan sonar: a practical guide", Int. Hyd. Rev., vol. LIII, no. 1, p. 65-92, 1976

Flewellen, C., N. Millard, I. Rouse; "TOBI - a vehicle for deep ocean survey", Electronics and Comm. Eng. J., p. 85-93, 1993

Gardner, J.V., M.E. Field, H. Lee, B.E. Edwards, D.G. Masson, N.H. Kenyon, R.B. Kidd; "Ground-truthing 6.5-kHz sidescan sonographs: What are we really imaging ?", J. Geophys. Re., vol. 96, no. B4, p. 5955-5974, 1991

Hayes, M.P., P.T. Gough; "Broad-band Synthetic Aperture Sonar", IEEE J. Oceanic Engineering, vol. 17, no. 1, p.80-94, 1992

Hughes Clarke, J.E., L.A. Mayer, N.C. Mitchell, A. Godin, G. Costello; "Processing and interpretation of 95-kHz backscatter data from shallow-water multibeam sonars", IEEE Oceans'93 Proc., vol. II, p. 437-442, 1983

Kleinrock, M.C.; "Overview of sidescan sonar sytems and processing", IEEE Proc., p.77-83, 1991

Kleinrock, M.C.; "Capabilities of some systems used to survey the deep-sea floor", CRC Handbook of Geophysical Exploration at Sea, Hard Minerals, p 36-90, R.A. Geyer (ed.), Boca Raton, FL: CRC Press, 1992

Kleinrock, M.C., R.N. Hey, A.E. Theberge Jr.; "Practical geological comparison of some seafloor survey instruments", Geophys. Res. Lett., vol. 19, no. 13, p. 1407-1410, 1992

Kosalos, J.G., D. Chayes; "SeaMARC - A portable system for ocean bottom imaging and charting", IEEE Oceans'83 Proc., p. 649-656, 1983

Loncarevic, B.D., B.M. Scherzinger; "Compensation of ship attitude for multibeam sonar surveys", Sea Technology, p. 10-15, 1994

Mitchell, N.C.; "Improving GLORIA images using Sea Beam data", J. Geophys. Res., vol. 96, no. B1, p. 337-351, 1991

Mitchell, N.C., M.L. Somers; "Quantitative backscatter measurements with a long-range sidescan sonar", IEEE J. Oceanic Engineering, vol. 14, no. 4, p. 368-374, 1989

Mitchell, N.C., E. McAllister, C. Flewellen, C.L. Walker; "Sidescan Sonar: A BRIDGE Workshop, 13/14 July 1993", BRIDGE Report no. 4, Natural Environment Research Council, UK, 1994 de Moustier, C.; "State of the art in swath bathymetry systems", Int. Hyd. Rev., vol. 65, no. 2, 1988

Musser, D.D.; "GPS/DGPS in Offshore Navigation, Positoning", Sea Technology, p. 61-66, 1992

Preston, J.M.; "Stability of towfish used as sonar platforms", IEEE Oceans'92 Proc., p. 888-893, 1992

Robertson, K.G.; "Deep Sea Navigation Techniques", Marine Geophysical Researches, vol. 12, p. 3-8, 1990

Somers, M.L.; "Sonar imaging of the seabed: Techniques, performance, applications", in *Acoustic Signal Prcessing for Ocean Exploration*, J.M.F. Moura and I.M.G. Lourtie (eds.), p. 355-369, Canadian Govt., 1993

Somers, M.L.; "Resolving the issue: a look at resolution and related topics in sonar", in *Man-made objects on the seafloor*, p. 41-58, Society for Underwater Technology, UK, 1995

Somers, M.L., A.R. Stubbs; "Sidescan sonar", IEE Proceedings, vol. 131, Part F, no. 3, p.243-256, 1984

Somers, M.L., Q.J. Huggett; "From GLORIA to GLORI-B", Sea Technology, p. 64-68, 1993

Stewart, W.K., D. Chu, S. Malik, S. Lerner, H. Singh; "Quantitative seafloor characterisation using a bathymetric sidescan sonar", IEEE J. Oceanic Eng., vol. 19, no. 4, p. 599-610, 1994

Telford, W.M., L.P. Geldart, R.E. Sheriff; "Applied Geophysics", 2nd edn, 770 pp., Cambridge University Press, UK, 1990

Wright, A.S.C.; "Deep-towed sidescan sonars", IEEE Oceans'93, vol. III, p. 478-483, 1993

Chapter 3: Sonar Data Processing

Bergersen, D.D.; "A synopsis of SeaMARC II sidescan processing techniques", Proc. IEEE Oceans'91, vol. 2, p. 921-926, 1991

Blondel, Ph.; "Traitement et Interprétation des Données Radar: Applications à l'étude de la surface de Vénus" (Treatment and Interpretation of Radar Data: Applications to the study of the surface of Venus), Ph.D. Thesis, Univ. Paris-VII Jussieu, 327 pp., 1992

Blondel, Ph., J.-C. Sempéré, V. Robigou, J.R. Delaney; " High-resolution bathymetry and geology of Endeavour Segment, Juan de Fuca Ridge", EOS Trans. AGU, 1993

Blondel, Ph., L.M. Parson; "Sonar Processing in the U.K.: A short review of existing potential and new developments for the BRIDGE Community", 27 pp., BRIDGE Publication no. 1, Natural Environment Research Council, UK, 1994

Blondel, Ph., L.M. Parson; "Sonar Processing in the UK", BRIDGE Report no. 5, 14 pp., Natural Environment Research Council, UK, 1995

Boyle, F.A., and N.P. Chatiros; "A model for acoustic backscatter from muddy sediments", Journal of the Acoustical Society of America, vol. 98, no. 1, p. 525-530, 1995

Burrough, P.A.; "Principles of Geographical Information Systems for land resources assessment", Oxford University Press, 194 pp., 1986

Caruthers, J.W., J.C. Novarini; "Estimating Geomorphology and Setting the Scale Partition with a Composite-Roughness Scattering Model", IEEE Oceans '93 Proc., vol. III, p. 220-228, 1993

Chavez, P.S.; "Processing techniques for digital sonar images from GLORIA", Photogrammetric Eng. and Remote Sens., vol. 52, no. 4, p. 365-388, 1986

Cobra, D.T.; "Estimation of geometric distortions in sidescan images", Proc. IEEE Oceans'91, vol. 2, p. 927-935, 1991

Dierckx, P.; "An algorithm for least-squares fitting of cubic spline surfaces to functions on a rectilinear mesh over a rectangle", J. Comp. Applied Math., vol. 3, p. 113-129, 1977

Fleming B.W.; "Sidescan sonar: a practical guide", Int. Hyd. Rev., vol. LIII, no. 1, p. 65-92, 1976

Gonzalez, R.C., P. Wintz; "Digital Image Processing", Reading, MA: Addison-Wesley, 431 pp., 1977

Hayes, M.P., P.T. Gough; "Broad-band synthetic aperture sonar", IEEE J. Oceanic Eng., vol. 17, no. 1, p. 80-94, 1992

Inoue, H.; "A least-squares smooth fitting for irregularly spaced data: finite-element approach using the cubic B-spline basis", Geophysics, vol. 51, no. 11, p. 2051-2066, 1986

Johnson, D.; "Sidescan sonar imagery analysis techniques", Proc. IEEE Oceans'91, vol. 2, p. 935-941, 1991

Johnson, H.P., M. Helferty; "The geological interpretation of side-scan sonar", Rev. of Geophysics, vol. 28, no. 4, p. 357-380, 1990

Kleinrock, M.C.; "Capabilities of some systems used to survey the deep-sea floor", CRC Handbook of Geophysical Exploration at Sea, Hard Minerals, p. 36-90, R.A. Geyer (ed.), Boca Raton, FL: CRC Press, 1992

LeBas, T.P., D.C. Mason, N.W. Millard; "TOBI Image Processing: The State of the Art", submitted to IEEE Oceanic Eng., 1994

Mason, D.C., T.P. LeBas, I. Sewell, C. Angelikaki; "Deblurring of GLORIA sidescan sonar images", Marine Geophys. Res., vol. 14, no. 2, p. 125-136, 1992

Matheron, G.: "La théorie des variables régionalisées et ses applications", CMM Techn. Rep., 1970

de Moustier, C.; "State of the art in swath bathymetry survey systems". Int. Hydr. Rev., vol. 65, no. 2, p. 25-54, 1988

Pouliquen, E.; "Identification des fonds marins superficiels à l'aide de signaux d'écho-sondeurs", Ph.D. Thesis, Univ. Paris-VII Jussieu, 204 pp., 1992

Preston, J.M.; "Stability of towfish used as sonar platforms", IEEE Proc. Oceans'92, p. 888-893, 1992

Reed,T.B., D. Hussong; "Digital Image Processing Techniques for Enhancement and Classification of SeaMARC II Side Scan Sonar Imagery", J. Geophys. Res., vol. 94, no. B6, p. 7469-7490, 1989

Searle, R.C., T.P. LeBas, N.C. Mitchell, M.L. Somers, L.M. Parson, Ph. Patriat, "GLORIA Image Processing: The State of the Art", Marine Geophys. Res., vol. 12, p. 21-39, 1990

Sempéré, J.-C., and EW9210 Scientific Party; "EW-9210 Cruise Report", unpub. report, 220 pp., 1992

Sempéré, J.-C., Ph. Blondel, A. Briais, T. Fujiwara, L. Géli, N. Isezaki, J.E. Pariso, L. Parson, Ph. Patriat, C. Rommevaux; "The Mid-Atlantic Ridge between 29°N and 31°30'N in the last 10 Ma", Earth and Planetary Science Letters, vol. 130, p. 45-55, 1995

Sharma, R.; "Recommendations for NASA's Space Physics Data System", EOS Trans. AGU, vol. 74, no. 50, p. 589-590, 1993

Smith, W.H.F., P. Wessel; "Gridding with continuous curvature splines in tension", Geophysics, vol. 55, no. 3, p. 293-305, 1990

Somers, M.L.; "Sonar imaging of the seabed", in Acoustic Signal Processing for Ocean Exploration, Moura and Lourtie (eds.), p. 355-369, 1993

Somers, M.L., Q.J. Huggett; " From GLORIA to GLORI-B", Sea Technology, p. 64-68, 1993

Stewart, W.K.S.; "Subsea Data Processing Standards", RIDGE Workshop Report, 1990

Ulaby, F.T., R.K. Moore, A.K. Fung; "Microwave remote sensing: active and passive", in Radar remote sensing and surface scattering and emission theory, vol. 2, Reading, MA: Addison-Wesley, 1982

Wessel, P., W.H.F. Smith; "Free software helps map and display data", EOS Trans. AGU, vol. 72, no. 441, 445-446, 1991

Chapter 4: Deep-Ocean Trenches and Collision Margins

An, L.-J., C.G. Sammis; "Development of strike-slip faults: shear experiments in granular materials and clay using a new technique", Journal of Structural Geology, 18(8), p. 1061-1077, 1996

Bangs, N.L.B., G.K. Westbrook, J.W. Ladd and P. Buhl, "Seismic velocities from the Barbados Ridge complex: indicators of high pore pressures in an accretionary complex", Journal of Geophysical Research, 95, 8767-8782, 1990

Brown, K., G K Westbrook; "Mud diapirism and subcretion in the Barbados Ridge Accretionary Complex: the role of fluids in accretionary processes", Tectonics, vol. 7, p. 613-640, 1988

Cande, S.C., R.B. Leslie; "Late Cenozoic tectonics of the southern Chile Trench", Journal of Geophysical Research, 91(B1), p. 471-496, 1986

Cande, S.C., R.B. Leslie, J.C. Parra, M. Hobart; "Interaction between the Chile Ridge and Chile Trench: geophysical and geothermal evidence", Journal of Geophysical Research, 92(B1), p. 495-520, 1987

Dickinson, W.R., D.R. Seely, "Structure and stratigraphy of forearc regions", Bulletin of the American Association of Petroleum Geologists, vol. 63, p. 2-31, 1979

Fitch, T.J.; "Plate convergence, transcurrent faults and internal deformation adjacent to southeast Asia and the western Pacific", Journal of Geophysical Research, vol. 77, p. 4432-4460, 1972

Fryer, P., H. Fryer; "Origins of non-volcanic seamounts in a forearc environment". In Seamounts, Islands and Atolls., B Keating, P. Fryer and R Batiza (eds), American Geophysical Union, AGU Monograph Series, 43, p. 61-69, 1977

Fryer, P., J.A. Pearce; "Introduction to the scientific results of Leg 125", in P. Fryer, J.A. Pearce and L.B. Stokking et al., Proc. ODP, Sci. Results, 125: College Station, TX (Ocean Drilling Programme), 1992

Henry, P.; "Fluid flow in and around a mud volcano field seaward of the Barbados accretionary wedge: results from Manon cruise", Journal of Geophysical Research, 101(B9), p. 20297-20323, 1990

Leggett J (ed.), "Trench-forearc geology: sedimentation and tectonics on modern and ancient active plate margins", Geol. Soc. Spec. Pub., 576 pp., 1982

Martinez, F., P. Fryer, N.A. Baker, T. Yamazaki; "Evolution of back-arc rifting: Mariana Trough, 20°-24°N", Journal of Geophysical Research, vol. 100, no. B3, p. 3807-3827, 1995

Moore, G.F., T.H. Shipley, P.L. Stoffa, D.E. Karig, A. Taira, S. Koramoto, H. Tokuyama, K. Suyehiro; "Structure of the Nankai Trough accretionary zone from multichannel seismic reflection data", Journal of Geophysical Research, 95, p. 8735-8765, 1990.

Pickering, K.T., M.B. Underwood, A. Taira, J. Ashi, "IZANAGI sidescan sonar and high-resolution multichannel seismic reflection line interpretation of accretionary prism and trench, offshore Japan". In "*Atlas of Deep Water Environments: Architectural style in turbidite systems*". K.T. Pickering, R.N. Hiscott, N.H. Kenyon, F.R. Lucchi, R.D.A. Smith (eds.), London: Chapman & Hall: London, p. 34-49, 1995

Schweller, W.J., L.D. Klum; "Depositional pattern and channelized sedimentation in active eastern Pacific trenches". In "*Sedimentation in submarine canyons, fans and trenches*", D.G. Stanley and G. Keling (eds), Strandsburg: Dowden, Hutchinson and Ross, p. 323-350, 1978

Thornburg, T.M., L.D. Kulm, D.M. Hussong; "Submarine-fan development in the southern Chile Trench: a dynamic interplay of tectonics and sedimentation", Geological Society of America Bulletin, 102(12), p. 1658-1680, 1990.

Westbrook, G.K., M.J. Smith; "Long decollements and mud volcanoes; evidence from the Barbados Ridge complex for the role of high pore fluid pressure in the development of an accretionary complex", Geology, vol. 11, p. 279-283, 1983

Westbrook, G.K., N. Hardy, and R Heath, "Structure and tectonics of the Panama-Nazca plate boundary". Geol. Soc. Am. Spec. Paper 295, 1995

Woodcock, N.H., M. Fischer; "Strike-slip duplexes", Journal of Structural Geology, vol. 7, p. 725-735, 1986

Chapter 5: Mid-Ocean Ridge Environments

Allerton, S., Murton, B.J., Searle, R.C., Jones, M.; "Extensional faulting and segmentation of the Mid-Atlantic Ridge north of the Kane Fracture Zone (24°00N to 24°40N)". Marine Geophysical Researches, 17(1), 37-61, 1995

Ballard, R.D., T.H. Van Andel; "Morphology and tectonics of the inner rift valley at latitude 36°50'N on the Mid-Atlantic Ridge", Geological Society of America, vol. 88, p. 507-530, 1977

Cann, J., C. Walker; "Breaking new ground on the ocean floor". New Scientist, 140 (1897), 24-29,1993

Cann, J., D.K. Smith, M.E. Dougherty, J. Lin, B. Brooks, S. Spencer, C.J. Macleod, E. Mcallister, R.A. Pascoe, J.A. Keeton; "Major landslides in the MAR median valley, 25-3°N: Their role in crustal construction and plutonic exposure". Eos: Transactions, American Geophysical Union, 73(43), Supplement, p. 569, 1992.

Cochran, J.R., J.A. Goff, A. Malinverno, D.J. Fornari, C. Keeley X Wang; "Morphology of a 'superfast' mid-ocean ridge crest and flanks: The East Pacific Rise, 7°-9°S", Marine Geophysical Researches, 15(1), p. 65-75, 1993

Detrick, R.S., S.E. Humphris; "Exploration of global oceanic ridge system unfolds". Eos: Transactions, American Geophysical Union, 75(29), p. 325-326, 1994

Frankel, H.; "From continental drift to plate tectonics", Nature, 335(6186), 127-130, 1988

Gente, P., J.-M. Auzende, V. Renard, Y. Fouquet, D. Bideau,; "Detailed geological mapping by submersible along the East Pacific Rise axial graben near 13°N". Earth and Planet Sci. Lett., 78, p. 224-226, 1986

Gente, P., R.A. Pockalny, C. Durand, C. Deplus, M. Maia, G. Ceuleneer, C. Mevel, M. Cannat, C. Laverne; "Characteristics and evolution of the segmentation of the Mid-Atlantic Ridge between 20°N and 24°N during the last 10 million years". Earth and Planetary Science Letters, 129, p. 55-71, 1995

German, C.R., L.M. Parson, R.A. Mills; "Mid-ocean ridges and hydrothermal activity", p. 152-164. In Oceanography: An Illustrated Guide (Ed. C.P. Summerhayes, S.A. Thorpe), London: Manson, 352 pp., 1996

Grindlay, N.R., P.J. Fox, P.R. Vogt; "Morphology and tectonics of the Mid-Atlantic Ridge (25°-27.30°S) from Seabeam and magnetic data", Journal of Geophysical Research, 97(B5), p. 6983-7010, 1992

Hayes, D.E., K.A. Kane; "Long-lived mid-ocean ridge segmentation of the Pacific-Antarctic Ridge and the Southeast Indian Ridge." Journal of Geophysical Research, 99(B10), p. 19679-19692,1994

Haymon, R.M., D.J. Fornari, K.L. Von Damm, M.D. Lilley, M.R. Perfit, J.M. Edmond, W.C. Shanks, R.A Lutz, J.M. Grebmeier, S. Carbotte, D. Wright, E. McLaughlin, M. Smith, N. Beedle, E. Olson; "Volcanic eruption of the mid-ocean ridge along the East Pacific Rise crest at 9.45-52°N: Direct submersible observations of seafloor phenomena associated with an eruption event in April, 1991", Earth and Planetary Science Letters, 119(1/2), p. 85-101, 1993

Karson, J.A.; "Tectonics of slow-spreading ridges". Oceanus, 34(4), p. 51-59, 1991/92, 1992

Karsten, J.A., J.R. Delaney, J.M. Rhodes, R.A. Liias; "Spatial and temporal evolution of magmatic systems beneath the Endeavour Segment, Juan de Fuca Ridge: Tectonic and petrologic constraints". Journal of Geophysical Research, 95(B12), p. 19235-19256, 1990

Kearey, P., F.J. Vine; "Global tectonics", 2nd edn., Oxford: Blackwell Science, 333 pp., 1996

Kleinrock, M.C., S.E. Humphris; "Structural control on sea-floor hydrothermal activity at the TAG active mound", Nature, 382(6587), p. 149-153, 1996.

Magde, L.S., D.K. Smith; "Seamount volcanism at the Reykjanes Ridge: Relationship to the Iceland hotspot", Journal of Geophysical Research, 100(B5), 8449-8468, 1995

Morgan, J.P.; "Mid-ocean ridge dynamics: Observations and theory", Reviews of Geophysics, 29, Supplement, Part 5, p. 807-822, 1995

Morgan, J.P., Y.J. Chen; "Dependence of ridge-axis morphology on magma supply and spreading rate", Nature, 364(6439), p. 706-708, 1993

Murton, B. J., L.M. Parson; "Segmentation, Volcanism and deformation of oblique spreading centres: a quantitative study of the Reykjanes Ridge", Tectonophysics, 222(2), p. 237-257, 1993

Needham, H.D., C.H. Langmuir (eds); "FARA-InterRidge Mid-Atlantic Ridge symposium: Results from 15°N-40°N"; Journal of Conference Abstracts, vol. 1, no. 2, p. 742-888, Cambridge Publications, 1996

Nicolas, A.; "The mid-oceanic ridges: Mountains below sea level." Berlin: Springer Verlag, 200 pp., 1995

Parson, L.M., J.W. Hawkins, P.M. Hunter; "Morphotectonics of the Lau Basin seafloor - Implications for the opening history of backarc basins", Proceedings of the Ocean Drilling Program, Initial Reports, 135 (Part 1), p. 81-82, 1992

Parson, L.M., B.J. Murton, R.C. Searle, D. Booth, J. Evans, P. Field, J.A. Keeton, A. Laughton, E. Mcallister, N. Millard, L. Redbourne, I. Rouse, A. Shor, D. Smith, S. Spencer, C. Summerhayes, C. Walker; "En echelon axial volcanic ridges at the Reykjanes Ridge: a life cycle of volcanism and tectonics", Earth and Planetary Science Letters, 117(1/2), p. 73-87, 1993

Parson, L.M., P. Patriat, R.C. Searle, A. Briais; "Segmentation of the Central Indian Ridge between 12°12'S and the Indian Ocean Triple Junction", Marine Geophysical Researches, 15(4), p. 265-282, 1993

Pezard, P.A., R.N. Anderson, W.B.F. Ryan, K. Becker, J.C. Alt, P. Gente; "Accretion, structure and hydrology of intermediate spreading-rate oceanic crusts from drillhole experiments and seafloor observations", Marine Geophysical Researches, 14(2), 93-123, 1992

Phipps Morgan, J.; "Mid-ocean ridge dynamics", IUGG National Report, 1987-1990, p. 807-822, American Geophysical Union, 1991

Ribe, N.M.; "On the dynamics of mid-ocean ridges", Journal of Geophysical Research, 93(B1), p. 429-436, 1988

Rona, P. A.; "Black smokers, massive sulphides and vent biota at the Mid-Atlantic Ridge", Nature, 321, p. 33-37, 1986

Searle, R.C.; "Tectonic pattern of the Azores spreading centre and Triple Junction", Earth and Planetary Science Letters, vol. 51, p. 415-434, 1980

Searle, R.C., A.S. Laughton; "Fine-scale sonar study of tectonics and volcanism on the Reykjanes Ridge", Oceanologica Acta, 4, p. 5-13, 1981

Seibold, E., W.H. Berger; "The sea floor: an introduction to marine geology", 2nd edn., Berlin: Springer Verlag, 356 pp., 1993

Sempéré, J.-C., J. Palmer, D.M. Christie, J.P. Morgan, A.N. Shor; "Australian-Antarctic Discordance", Geology, 19(5), p. 429-432, 1991

Sempéré, J.-C., Ph. Blondel, A. Briais, T. Fujiwara, L. Géli, N. Isezaki, J.E. Pariso, L. Parson, Ph. Patriat, C. Rommevaux; "The Mid-Atlantic Ridge between 29°N and 31°30'N in the last 10 Ma", Earth and Planetary Science Letters, vol. 130, p. 45-55, 1995

Sinton, J.M.; "Evolution of mid ocean ridges". Washington, DC: American Geophysical Union, 77 pp. (Geophysical Monograph 57) (IUGG vol. 8), 1989

Smith, D.K., J.R. Cann; "Hundreds of small volcanoes on the median valley floor of the Mid-Atlantic Ridge at 24-30°N", Nature, 348(6297), p. 152-155, 1990

Smith, D.K., J.R. Cann; "The role of seamount volcanism in crustal construction at the Mid-Atlantic Ridge (24°-30°N)", Journal of Geophysical Research, 97(B2), p. 1645-1658, 1992

Smith, D.K., J.R. Cann, M.E. Dougherty, J. Lin, S. Spencer, C.J. Macleod, J.A. Keeton, E. Mcallister, B. Brooks, R. Pascoe, W. Robertson; "Mid-Atlantic Ridge volcanism from deep-towed side-scan sonar images, 25-29°N". Journal of Volcanology and Geothermal Research, 67, p. 233-262, 1995

Smith, D.K., S.E. Humphris, W.B. Bryan; "A comparison of volcanic edifices at the Reykjanes Ridge and the Mid-Atlantic Ridge at 24-30°N", Journal of Geophysical Research, 100(B11), p. 22485-22498, 1995

Taylor, R.N., Murton, B.J., M.F. Thirlwall; "Petrogenesis and geochemical variation along the Reykjanes Ridge, 57°N-59°N", Journal Geol. Soc. London, 152, p. 1031-1037, 1995

Tharp, M.; "Discovery of the mid-ocean rift system", Lamont-Doherty Geological Observatory of Columbia University, Annual Report, 1989, p. 20-21, 1989

Tucholke, B.E.; "Massive submarine rockslide in the rift-valley wall of the Mid-Atlantic Ridge", Geology, 20(2), p. 129-132, 1992

Tunnicliffe, V.; "The Biology of Hydrothermal Vents: Ecology and Evolution". Oceanography and Marine Biology Annual Review, 29, p. 319-407, 1991

Van Dover, C. L.; "Ecology of Mid-Atlantic Ridge hydrothermal vents". In Parson, L.M., C.L. Walker, D.R. Dixon (eds), *Hydrothermal vents and processes*, Geol. Soc. Spec. Pub. no. 87, p. 257-294, 1995

Vine, F. J., D.H. Matthews;."Magnetic anomalies over oceanic ridges", Nature, 199, p. 947-949, 1963

Weiland, C.M., K.C. Macdonald, N.R. Grindlay; "Ridge segmentation and the magnetic structure of the southern Mid-Atlantic Ridge, 26°S and 31°35'S: Implications for magmatic processes at slow spreading centers", Journal of Geophysical Research, 101(B4), p. 8055-8073, 1996

Wolfe, C.J., G.M. Purdy, D. Toomey, S.C. Solomon; "Microearthquake characteristics and crustal velocity structure at 29N on the Mid-Atlantic Ridge: The architecture of a slow spreading segment", Journal of Geophysical Research, 100(B12), p. 24449-24472, 1995

Chapter 6: Abyssal Plains and Basins

Angel, M.V.; "Deep abyssal plains: do they offer a viable option for the disposal of large-bulk low-toxicity wastes ?", p. 61-75, in *Advances in the science and technology of ocean management*, H.D. Smith (ed), London: Routledge, 240 pp., 1992

Faugères, J.-C., D.A.V. Stow; "Bottom-current-controlled sedimentation: a synthesis of the contourite problem", Sedimentary Geology, vol. 82, p. 287-297, 1993

Holcomb, R.T., R.C. Searle; "Large landslides from oceanic volcanoes", Marine Geotechnology, vol. 10, p. 19-32, 1991

Hollister, C.D.; "The concept of deep-sea contourites", Sedimentary Geology, vol. 82, p. 5-11, 1993

Jacobs, C.; "Mass wasting along the Hawaiian Ridge: giant debris avalanches", p. 26-28, in *Atlas of Deep Water Environments: Architectural style in turbidite systems*, K.T. Pickering, R.N. Hiscott, N.H. Kenyon, F. Ricci Lucchi, R.D.A. Smith (eds), London: Chapman & Hall, 333 pp., 1995

Kidd, R.B.; "Long-range sidescan sonar study of sediment slides and the effect of slope mass sediment movement on abyssal plain sedimentation", p. 289-303 in *Marine slides and other mass movements*, S. Saxov, J.K.Nieuwenhuis (eds), New York: Plenum, 353 pp., 1982

Kidd, R.B., R.W. Simm, R.C. Searle; "Sonar acoustic facies and sediment distribution on an area of the deep ocean floor", Marine and Petroleum Geology, 2(3), 210-221, 1985

Masson, D.G.; "Late Quaternary turbidity current pathways to the Madeira Abyssal Plain and some constraints on turbidity current mechanisms", Basin Research, vol. 6, no. 1, p. 17-33, 1994

Masson, D.G.; "Catastrophic collapse of the volcanic island of Hierro 15 ka ago and the history of landslides in the Canary Islands", Geology, vol. 24, no. 3, p. 231-234, 1996

Masson, D.G., Q.J. Huggett, P.P.E. Weaver, D. Brunsden, R.B. Kidd; "The Saharan and Canary debris flows offshore Northwest Africa", Landslide News, Tokyo, no. 6, p. 9-13, 1992

Masson, D.G., R.B. Kidd, J.V. Gardner, Q.J. Huggett, P.P.E. Weaver; "Saharan continental rise: facies distribution and sediment slides", p. 327-343 , in Poag, C.W., P.C. de Graciansky; Geologic Evolution of Atlantic Continental Rises, New York: Van Nostrand Reinhold, 378 pp., 1992

Masson, D.G., Q.J. Huggett, D. Brunsden; "The surface texture of the Saharan Debris Flow deposit and some speculations on submarine debris flow processes", Sedimentology, vol. 40, p. 583-598, 1993

Moore, J.G., G.W. Moore; "Deposit from a giant wave on the island of Lanai, Hawaii", Science, vol. 226, p. 1312-1315, 1984

Moore, J.G., D.A. Clague, R.T. Holcomb, P.W. Lipman, W.R. Normark, M.E. Torresan; "Prodigious submarine landslides on the Hawaiian Ridge", Journal of Geophysical Research, vol. 94, no. B-12, p. 17,465-17,484, 1989

Müller, P., G. Holloway; "Workshop revisits topographic effects in the oceans", EOS Trans. AGU, p. 300-303, vol. 76, no. 30, 1995

O'Leary, D.W., M.R. Dobson; "Southeastern New England continental rise: Origin and history of slide complexes", p. 214-265 , in Poag, C.W., P.C. de Graciansky; Geologic Evolution of Atlantic Continental Rises, New York: Van Nostrand Reinhold, 378 pp. 1992

Pickering, K.T., R.N. Hiscott, N.H. Kenyon, F. Ricci Lucchi, R.D.A. Smith; "Atlas of Deep Water Environments: Architectural style in turbidite systems", London: Chapman & Hall, 333 pp., 1995

Pilkey, O.H.; "Sedimentology of basin plains", in Geology and Geochemistry of Abyssal Plains, P.P.E. Weaver, J. Thomson (eds), Oxford: Blackwell for the Geological Society, 246 pp., 1987

Pratson, L.F., E.P. Laine; "The relative importance of gravity-induced versus current-controlled sedimentation during the Quaternary along the middle U.S. continental margin revealed by 3.5 kHz echo character", Marine Geology, vol. 89, p. 87-126, 1989

Price, N. J., J. W. Cosgrove; "Analysis of geological structures", Cambridge University Press, 502 pp., 1990

Schuttenhelm, R.T.E., G.A. Auffret, D.E. Buckley, R.E. Cranston, C.N. Murray, L.E. Shepard, A.E. Spijkstra, "Geoscience investigations of two North Atlantic abyssal plains: the ESOPE international expedition", 2 vols, 1293 pp., Luxembourg: Office of Official Publications of the European Communities, 1989

Searle, R.C. P.J. Schultheiss, P.P.E. Weaver, M. Noel, R.B. Kidd, C.L. Jacobs, Q.J. Huggett; "Great Meteor East (distal Madeira Abyssal Plain): geological studies of its suitability for disposal of heat-emitting radioactive wastes", Institute of Oceanographic Sciences, Report no. 193, 162 pp., 1985

Telford, W.M., L.P. Geldart, R.E. Sheriff; "Applied Geophysics", 2nd edn, Cambridge University Press, 770 pp., 1990

Weaver, P.P.E., J. Thomson; "Geology and Geochemistry of Abyssal Plains", Oxford: Blackwell for the Geological Society, 246 pp., 1987

Weaver, P.P.E., R.G. Rothwell, J. Ebbing, D. Gunn, P.M. Hunter; " Correlation, frequency of emplacement and source directions of megaturbidites on the Madeira Abyssal Plain", Marine Geology, vol. 109, no. 1/2, p. 1-20, 1992

Weaver, P.P.E., D.G. Masson, D.E. Gunn, R.B. Kidd, R.G. Rothwell, D.A. Maddison; "Sediment mass wasting in the Canary Basin", p. 287-296, in *Atlas of Deep Water Environments: Architectural style in turbidite systems*, K.T. Pickering, R.N. Hiscott, N.H. Kenyon, F. Ricci Lucchi, R.D.A. Smith (eds), London: Chapman & Hall, 333 pp., 1995

Chapter 7: Continental Margins

Almagor, G., G. Wiseman; "Analysis of submarine slumping in the continental slope off the southern coast of Israel", Marine Geotechnology, 10(3/4), 303-342, 1991

Behrens, E.W.; " Geology of a continental slope oil seep, Northern Gulf of Mexico", American Association of Petroleum Geologists Bulletin, 72(2), 105-114, 1988

Beijdorff, C., W. van der Werff, Yu. Gubanov; "Eratosthenes Seamount; MAK-1 sonographs and profiles", p. 129-133, in *Mud Volcanism in the Mediterranean and Black Seas and shallow structure of the Eratosthenes Seamount*, A.F. Limonov, J.M. Woodside, M.K. Ivanov (eds), UNESCO Reports in Marine Science no. 64, 173 pp., 1994

Booth, J.S., D.W. O'Leary; "A statistical overview of mass movement characteristics on the North American Atlantic outer continental margin", Marine Geotechnology, 10(1/2), 1-18, 1991

Bouma, A.H., H.H. Roberts (eds), "Northern Gulf of Mexico continental slope", Geo-Marine Letters, vol., no. 4, Special Issue, 1990

Carlson, P.R. H.A. Karl, B.D. Edwards; "Mass sediment failure and transport features revealed by acoustic techniques, Beringian Margin, Bering Sea, Alaska", Marine Geotechnology, 10(1/2), 33-51, 1991

Chauhan, O.S., F. Almeida, C. Moraes; "Regional geomorphology of the continental slope of NW India: delineation of the signatures of deep-seated structures", Marine Geodesy, 15(4), 283-296, 1992

Chough, S.K. R. Hesse; "Contourites from Eirik Ridge, south of Greenland", Sedimentary Geology, vol. 41, p. 185-199, 1985

Chough, S.K., H.J. Lee; "Submarine slides in the eastern continental margin, Korea", Marine Geotechnology, 10(1/2), 71-82, 1991

Cita, M.B., M.K. Ivanov, J.M. Woodside (eds), "The Mediterranean Ridge diapiric belt", Marine Geology, vol. 132, Special Issue, 1996

Cochonat, P., L. Droz, C. Geronimi, J. Guillaume, B. Loubrieu, G. Ollier, J.-P. Peyronnet, A. Robin, R. Tofani; "Submarine morphology of the eastern part of the Niger delta (Gulf of Guinea)", C.R. Acad. Sci. Paris, Serie II, 317, 1317-1328, 1993

Coussot, P., M. Meunier; "Recognition, classification and mechanical description of debris flows", Earth Science Reviews, vol. 40, no. 3/4, p. 209-227, 1996

Dorn, W.U., F. Werner; "The contour-current flow along the southern Iceland-Faeroe Ridge as documented by its bedforms and asymmetrical channel fillings", Sedimentary Geology, vol. 82, p. 47-59, 1993

EEZ-SCAN 87 Scientific Staff "Atlas of the U.S. Exclusive Economic Zone: Atlantic continental margin", United States Geological Survey, Miscellaneous Investigations Series, vol. I-2054, 174 pp., 1991

Farre, J.A., W.B.F. Ryan; "3-D view of erosional scars on U.S. Mid-Atlantic continental margin", American Association of Petroleum Geologists Bulletin, 69(6), 923-932, 1985

Fusi, N. N.H. Kenyon; "Distribution of mud diapirism and other geological structures from long-range sidescan sonar (GLORIA) data, in the Eastern Mediterranean Sea", Marine Geology, vol. 132; p. 21-38, 1996

Galindo-Zaldivar, J., L. Nieto, J.M. Woodside; "Structural features of mud volcanoes and the fold system of the Mediterranean Ridge, south of Crete", Marine Geology, vol. 132; p. 95-112, 1996

Gardner, J.V., R.B. Kidd; "Sedimentary processes on the northwestern Iberian continental margin viewed by long-range side-scan sonar and seismic data", Journal of Sedimentary Petrology, 57(3), 397-407, 1987

Ginsburg, G.D., V.A. Soloviev, R.E. Cranston, T.D. Lorenson, K.A. Kvenvolden; "Gas hydrates from the continental slope, offshore Sakhalin Island, Okhotsk Sea", Geo-Marine Letters, 13(1), 41-48, 1993

Hagen, R.A., D.D. Bergersen, R. Moberly, W.T. Coulbourn; "Morphology of a large meandering submarine canyon system on the Peru-Chile forearc", Marine Geology, vol. 119, p. 7-38, 1994

Hampton, M.A., H.A. Karl, N.H. Kenyon; "Sea-floor drainage features of Cascadia Basin and the adjacent continental slope, northeast Pacific Ocean", Marine Geology, 87(2/4), 249-272, 1989

Hieke, W., F. Werner; H.-W. Schenke; "Geomorphological study of an area with mud diapirs south of Crete (Mediterranean Ridge), Marine Geology, vol. 132, p. 63-93, 1996

Hiscott, R.N., A.E. Aksu; "Submarine debris flows and continental slope evolution in front of Quaternary ice sheets, Baffin Bay, Canadian Arctic", American Association of Petroleum Geologists Bulletin, 78(3), 445-460, 1994

Hovland, M. A.G. Judd; "Seabed pockmarks and seepages: impact on geology, biology and the marine environment", London: Graham & Trotman, 293 pp., 1988

Ivanov, M.K., A.F. Limonov, Tj.C.E. van Weering; "Comparative characteristics of the Black Sea and Mediterranean Ridge mud volcanoes", Marine Geology, vol. 132, p. 253-271, 1996

Jenkins, C.J., J.B. Keene; "Submarine slope failures of the southeast Australian continental slope: a thinly sedimented margin", Deep-Sea Research, 39A(2), 121-136, 1992

Jenkins, C.J., B. Hunt, M. Lawrence; "GLORIA super sidescan imagery of the continental slope east of Sydney - Batemans Bay", Hydrographic Society Special Publication, no.27, 291-298, 1991

Kenyon, N.H.; "Evidence from bedforms for a strong poleward current along the upper continental slope of NW Europe", Marine Geology, vol. 72, p. 187-198, 1986

Kenyon, N.H.; "Mass-wasting features on the continental slope of northwest Europe", Marine Geology, 74(1/2), 57-77, 1987

Kenyon, N.H., R.H. Belderson, A.H. Stride; "Channels, canyons and slump folds on the continental slope between south-west Ireland and Spain", Oceanologica Acta, 1, 369-380, 1978

Kidd, R.B. R.W. Simm, R.C. Searle; "Sonar acoustic facies and sediment distribution on an area of the deep ocean floor", Marine and Petroleum Geology, 2(3), 210-221, 1985

Knight, R.J., J.R. McLean (eds); "Shelf sands and sandstones", Canadian Society of Petroleum Geologists, Memoir II, , 1986

Limonov, A.F., J.M. Woodside, M.K. Ivanov (eds); "Mud volcanism in the Mediterranean and Black Seas and shallow structure of the Eratosthenes Seamount", UNESCO Reports in Marine Science no. 64, 173 pp., UNESCO, 1994

Limonov, A.F., J.M. Woodside, M.B. Cita, M.K. Ivanov; "The Mediterranean Ridge and related mud diapirism: a background", Marine Geology, vol. 132, p. 7-19, 1996

Milliman, J.D., W.R. Wright (eds); "The marine environment of the U.S. Atlantic continental slope and rise", Woods Hole, MA: Jones and Bartlett, 275pp., 1987

Neurauter, T.W., H.H. Roberts; "Three generations of mud volcanoes on the Louisiana continental slope", Geo-Marine Letters, 14(2/3), 120-125, 1994

Neurauter, T.W., W.R. Ryant; "Gas hydrates and their association with mud diapir/mud volcanoes on the Louisiana continental slope", in Offshore Technology Conference, OTC 89 Proceedings, vol. 1, p. 599-607, 1989

Palanques, A., N.H. Kenyon, B. Alonso, A. Limonov; "Erosional and depositional patterns in the Valencia Channel Mouth: An example of a modern channel-lobe transition zone", Marine Geophysical Researches, vol. 17, p. 503-517, 1995

Poag, C. W., P.C. de Graciansky; "Geologic evolution of Atlantic continental rises", 369 pp., New York: Van Nostrand Reinhold, 378 pp., 1992

Prior, D.B., E.H. Doyle; "Intra-slope canyon morphology and its modification by rockfall processes, U.S. Atlantic continental margin", Marine Geology, 67(1/2), 177-196, 1985

Rao, V.P. M. Veerayya, R.R. Nair, P.A. Dupeuble, M. Lamboy; "Late Quaternary Halimeda bioherms and aragonitic faecal pellet-dominated sediments on the carbonate platform of the western continental shelf of India", Marine Geology, vol. 121, p. 293-315, 1994

Roberts, H.H., C.V. Phipps, L.L. Effendi; "Morphology of large Halimeda bioherms, Eastern Java Sea (Indonesia): a side-scan sonar study", Geo-Marine Letters, vol. 7, no. 1, p. 7-14, 1987

Rothwell, R.G., N.H. Kenyon, B.A. McGregor; "Sedimentary features of the south Texas continental slope as revealed by side-scan sonar and high-resolution seismic data", American Association of Petroleum Geologists Bulletin, 75(2), 298-312, 1991

Saxov, S., J.K. Nieuwenhuis; "Marine slides and other mass movements", Plenum Press, New York, 1982

Tomlinson, J.S., C.J. Pudsey, R.A. Livermore, R.D. Larter, P.F. Barker; "Long-range sidescan sonar (GLORIA) survey of the Antarctic peninsular Pacific margin", in Recent Progress in Antarctic Science, Y.Yoshida et al. (eds), p. 423-429, Tokyo: Terra Scientific Publishing Company, 1992

Twichell, D.C., D.G. Roberts; "Morphology, distribution and development of submarine canyons on the United States Atlantic continental slope between Hudson and Baltimore Canyons", Geology, 10, 408-412, 1982

Volgin, A.V., J.M. Woodside; "Sidescan sonar images of mud volcanoes from the Mediterranean Ridge: possible causes of variations in backscatter intensity", Marine Geology, vol. 132, p. 39-53, 1996

Westbrook, G.K., and the MEDRIFF Consortium; "Three brine lakes discovered in the seafloor of the Eastern Mediterranean", EOS Trans. AGU, vol. 76, no. 33, p. 313-318, 1995

Westbrook, G.K. et al.; "Brine lakes and thermal anomalies in the sea floor and bottom water of the western Mediterranean Ridge", p.370-383 in *Marine Sciences and Technologies. Second MAST days and Euromar market, 7-10 November 1995*, Project reports Volume 1, M.Weydert, E.Lipiatou, R.Goni, C.Fragakis, M.Bohle-Carbonell, K.-G.Barthel (ed). Luxembourg: Office for Official Publications of the European Communities, 788pp., 1995

Woodside, J.M., A.V. Volgin; "Brine pools associated with Mediterranean Ridge mud diapirs: an interpretation of echo-free patches in deep-tow sidescan sonar data", Marine Geology, vol. 132, p. 55-61, 1996

Chapter 8: Coastal Environments

Allen, J.L.R.; "Sedimentary structures: their character and physical basis", Developments in Sedimentology, vol. II, Amsterdam: Elsevier, 663 p., 1984

Arntz, W.E., W. Weber; "*Cyprina islandica* L. (Mollusca, Bivalvia) als Nahrung für Dorsch und Kliesche in der Kieler Bucht", Ber. Dt. Wiss. Komm. Meeresforschung, 21, 193-209, 1970

Belderson, R.H., N.H.Kenyon, A.H. Stride, A.R. Stubbs; "Sonographs of the seafloor", Amsterdam: Elsevier, 185 pp., 1972

Belderson, R.H., M.A. Johnson, N.H. Kenyon; "Bedforms", Offshore Tidal Sands, Process and Deposits, A.H. Stride (ed), p. 27-57, 1982

Bernhard, M.; "Sedimentologische Beeinflussung der Oberflächensedimente durch Grundfischerei in der Kieler Bucht", Unveröff. Dipl. - Arbeit, Geol. Inst. Univ. Kiel, Germany, 41 pp., 1989

de Groot, S.J.; "The impact of bottom trawling on benthic fauna of the North Sea". Ocean Management, vol. 9, p. 177-199, 1984

Elverhoi, A.; "Glacigenic and associated marin sediments in the Weddell Sea, fjords of Spitsbergen and the Barents Sea: a review", Marine Geology, 57, 53-88, 1984

Fish, J.P., H.A. Carr; "Sound underwater images", Orleans, MA: Lower Cape Publishing, 189 pp., 1990

Flemming, B.W.; "Side-scan sonar: a practical guide", International Hydrographic Review, 53 (1), p. 65-91, 1976

Freiwald, A.; "R/V Victor Hensen (cruise 24/95): Deep-water coral reef mounds on the Sula-Ridge, Mid-Norwegian Shelf", unpubl. cruise report, Univ. Bremen, 13pp., 1995

Freiwald, A., J.B. Wilson, M.E. Willemson, V. Hühnerbach; "Giant deep-water Lophelia mounds on the Mid-Norwegian Shelf", Sedimentary Geology, 1997

Hambrey, M.; "Glacial environments", London: UCL Press Limited, 296 pp., 1994

Hennings, I., H. Pasenau, F. Werner; "Sea surface signatures related to subaqueous dunes detected by acoustic and radar sensors", Continental Shelf Research, vol. 13, no. 8/9, p. 1023-1043, 1993

Henrich, R., A. Freiwald, C. Betzler, B. Bader, P. Schäfer, C. Samtleben, T.C. Brachert, A. Wehrmann, H. Zankl, D.H.H. Kühlmann; "Controls on modern carbonate sedimentation on warm-temperate to arctic coasts, shelves and seamounts in the northern Hemisphere: implications for fossil counterparts", Facies, vol. 32, p. 71-108, 1995

Hovland, M. A.G. Judd; "Seabed pockmarks and seepages: impact on geology, biology and the marine environment", London: Graham & Trotman, 293 pp., 1988

Hühnerbach, V.; "Hierro debris avalanche and the upper part of the Canary Slide: Sidescan sonar imaging and its geological interpretation", unpublished report, University of Kiel, 69 pp., 1996

Jordan, G.F.; "Large sand waves in estuaries and in the open sea", Proc. First Nat. Coastal Shallow Water Res. Conf., p. 232-236, 1962

Kenyon, N.H.; "Sand ribbons of European tidal seas", Marine Geology, vol. 9, p. 25-39, 1970

Khandriche, A., F. Werner; "Freshwater-induced pockmarks in Bay of Eckernförde, Western Baltic", Prace Panstwowoego Instyitutu Geologicznego CXLIX, Proceedings of the Third Marine Geological Conference "The Baltic", J.E. Mojski (ed), p. 155-164, 1995

Khandriche, A., F. Werner, H. Erlenkeuser, H.; "Auswirkungen der Ostwindstürme vom Winter 1978/79 auf die Sedimentation im Schlickbereich der Eckernförder Bucht (Westliche Ostsee)", Meyniana, vol. 38, p. 125-152, 1986

King, L.H., K. Rokoengin, G.B.J. Fader, T. Gunleikrud; "Till tongue stratigraphy". Geological Society of America Bulletin, vol. 103, p. 637-659, 1991

Krost, P., M. Bernhard, F. Werner, W. Hukriede; "Otter trawl tracks in Kiel Bay (Western Baltic) mapped by side-scan sonar", Meeresforschungen, vol. 32, p. 344-353, 1990

Lien, R., A. Solheim, A. Elverhoi, K. Rokoengin; "Iceberg scouring and seabed morphology on the eastern Weddell Sea Shelf", Polar Research, vol. 4, p. 43-57, 1989

McCave, I.N.; "Recent shelf clastic sediments". In *Sedimentology: Recent Developments and applied aspects* , P.J. Brenchley, B.P.J. Williams (ed), Geological Society of London Special Publications, no. 18, p. 49-65, 1985

Newton, R.S., E. Seibold, F. Werner; "Facies distribution patterns on the Spanish Sahara continental shelf mapped with sidescan sonar", Meteor Forsch. Ergeb., C15, p. 55-77, 1973

Newton, R.S., A. Stefanon; "Application of sidescan sonar in marine biology", Marine Biology, vol. 31, p. 287-291, 1975

Off, T.; "Rhythmic linear sand bodies caused by tidal currents". Bull. Am. Assoc. Petrol. Geol., 47, p. 324-341, 1963

Parson, L.M., Ph. Blondel, T.P. LeBas; "Investigation of the sidescan data from the M/V Derbyshire wrecksite - Final Report", IOSDL Report, Southampton Oceanography Centre, UK, 1995

Reading, H.G.; "Sedimentary Environments and Facies", Oxford: Blackwell, 615 pp., 1978

Riesen, W., K. Reise; "Macrobenthos of the subtidal Wadden Sea revisited after 55 years", Helgoländer Meeresuntersuchungen, vol. 35, p. 409-423, 1982

Reineck, H.-E.; "Sedimentgefüge im Bereich der südlichen Nordsee", Abh. senckenbergische naturforsch. Ges., no. 505, 138 pp., 1963

Reineck, H.-E., I.B. Singh; "Depositional Sedimentary Environments", New York: Springer Verlag, 551 pp., 1980

Schacht, R., "Sonographische und sedimentologische Untersuchungen zur Klimaentwicklung in hocharktischen Fjordsystemen", unpubl. thesis, Univ. Kiel., 120 pp., 1996

Schröder, H.G.; "Sedimentoberflächen im östlichen Bodensee-Obersee- Sidescan Untersuchungen im Zusammenhang mit den Auswirkungen der Vorstreckung des Alpenrheins", Ber. Nr. 43 der Internationale Gewässerschutzkommision für den Bodensee, 48 pp., 1992

Schwarzer, K; "Auswirkungen der Sandentnahme auf dem Salzsand auf die morphologisch/sedimentologischen Verhältnisse", Unveröffent. Bericht, Amt f. Land- und Wasserwirtschaft, Husum, 11 pp., 1996

Schwarzer, K., K. Ricklefs, W. Schumacher, R. Atzler; "Beobachtungen zur Vorstanddynamik und zum Küstenschutz sowie zum Sturmereignis vom 3./4.11.1995 vor dem Streckelsberg/Usedom", Meyniana, vol. 48, p.49-68, 1996

Solheim, A.; "The depositional environment of surging sub-polar tidewater glaciers", Skrifter Norsk Polarinstitut, 194 pp, 1991

Stefanon, A.; "Marine sedimentology through modern acoustical methods: 1. Sidescan sonar", Bolletino di Oceanologia Teorica et Applicata, 3 (1), p. 3-38, 1985

Stride, A.H.; "Shape and size trends for sand waves in a depositional zone of the North Sea", Geol. Mag., p. 469-477, 1970

Terwindt, J.H.J.; "Lithofacies of inshore estuarine and tidal-inlet deposits", Geol. Mijnbouw, 50, p. 515-526, 1971

Ulrich, J.; "Untersuchungen zur Pendelbewegung von Tiderippeln im Heppenser Fahrwasser (Innenjade)", Die Küste, vol. 23, p. 112-121, 1972

Werner, F.; "Principios de Interpretacion del Sonar Lateral y Ejemplos de su Aplicacion en la Plataforma Continental Argentina", Instituto Argentino de Oceanografia, Universidad Nacional del Sur, Bahia Blanca, no. 38, 36 pp., 1977

Werner, F., R.S. Newton; "The pattern of large-scale bed forms in the Langeland Belt (Baltic Sea)", Marine Geology, vol. 19, p. 39-62, 1975

Werner, F., G. Unsöld, B. Koopmann, A. Stefanon; "Field observations and flume experiments on the nature of comet marks", Sedimentary Geology, vol. 26, p. 233-262, 1980

Werner, F., G. Hoffmann, M. Bernhard, D. Milkert, K. Vikgren; "Sedimentologische Auswirkungen der Grundfischerei in der Kieler Bucht", Meyniana, vol. 42, p. 123-151, 1990

Chapter 9: Image Anomalies and Sonar System Artefacts

Cervenka, P., C. deMoustier, P.F. Lonsdale; "Geometric corrections on sidescan sonar images based on bathymetry: Application with SeaMARC II and Sea Beam data", Marine Geophysical Researches, 16(5), 365-383, 1994

Cobra, D.T., A.V. Oppenheim, J.S. Jaffe; "Geometric distortions in side-scan sonar images: a procedure for their estimation and correction", IEEE Journal of Oceanic Engineering, 17(3), 252-268, 1992

Douglas, B.L., H. Lee, "Motion compensation for improved sidescan sonar imaging", pp. 378-383 in Oceans '93, vol. I, New York: IEEE, 491 pp., 1993

Elachi, C., L.E. Roth, G.G. Schaber; "Spaceborne radar subsurface imaging in hyperarid regions", IEEE Trans. Geoscience and Remote Sensing, vol. GE-22, p. 383-388, 1984

Fish, J.P., H.A. Carr; "Sound underwater images: a guide to the generation and interpretation of sidescan sonar data", EG&G Marine Instruments: Cataumet, 189 pp., 1990

Fleming B.W.; "Sidescan sonar: a practical guide", Int. Hyd. Rev., vol. LIII, no. 1, p. 65-92, January 1976

Fox, C.G., F.J. Jones, T.-K. Lau; "Constrained iterative deconvolution applied to SeaMARC I sidescan sonar imagery", IEEE Journal of Oceanic Engineering, 15(1), 24-31, 1990

Huggett, Q.J., A.K. Cooper, M.L. Somers, R.A. Stubbs; "Interference fringes on GLORIA sidescan sonar images from the Bering Sea and their implications", Marine Geophysical Researches, vol. 14, p. 47-63, 1992

Ivanov, M.K., A.F. Limonov, Tj.C.E. van Weering; "Comparative characteristics of the Black Sea and Mediterranean Ridge mud volcanoes", Marine Geology, vol. 132, p. 253-272, 1996

Johnson, D.; "Sidescan sonar imagery analysis techniques", Proc. IEEE Oceans'91, vol. 2, p. 935-941, 1991

Johnson, H.P., M. Helferty; "The geological interpretation of side-scan sonar", Rev. of Geophysics, vol. 28, no. 4, p. 357-380, 1990

LeBas, T.P., D.C. Mason; "Suppression of multiple reflections in GLORIA sidescan sonar imagery", Geophysical Research Letters, vol. 21, no. 7, p. 549-552, 1994

Li, R., "Correction of pixel locations of sidescan sonar images using bathymetric data acquired separately", Marine Geodesy, 15(2/3), 211-213, 1992

Mason, D.C., T.P. LeBas, I. Sewell, C. Angelikaki; "Deblurring of GLORIA side-scan sonar images", Marine Geophysical Researches, vol. 14, p. 125-136, 1992

Preston, J.M.; "Stability of towfish used as sonar platforms", IEEE Oceans'92 Proc., p. 888-893, 1992

Reed, T.B., D. Hussong; "Digital Image Processing Techniques for Enhancement and Classification of SeaMARC II Side Scan Sonar Imagery", Journal of Geophysical Research, vol. 94, no. B6, p. 7469-7490, 1989

Somers, M.L.; "Sonar imaging of the seabed: Techniques, performance, applications", in *Acoustic Signal Processing for Ocean Exploration*, J.M.F. Moura and I.M.G. Lourtie (eds), p. 355-369, Canadian Govt., 1993

Somers, M.L., A.R. Stubbs; "Sidescan sonar", IEE Proceedings, vol. 131, Part F, no. 3, p. 243-256, June 1984

Somers, M.L.; "Resolving the issue: a look at resolution and related topics in sonar", in *Man-made objects on the seafloor*, p. 41-58, Society for Underwater Technology, UK, 1995

Thorpe, S.A. A.R. Stubbs, A.J. Hall, R.J. Turner; "Wave-produced bubbles observed by side-scan sonar", Nature, vol. 296, p. 636-638, 1982

Voulgaris, G., M.B. Collins; "Linear features on side-scan sonar images: an algorithm for the correction of angular distortion", Marine Geology, 96(1/2), 187-190, 1991

Chapter 10: Computer-Assisted Interpretation

Ashkar, G. P., J. W. Modestino; "The contour extraction problem with biomedical applications", Computer Graphics and Image Processing, vol. 7, p. 331-355, 1978

Ballard, D. H.; "Generalizing the Hough transform to detect arbitrary shapes", IEEE Trans. Pattern Recognition, vol. 13, p. 111-122, 1981

Blondel, Ph., J.-C. Sempéré, V. Robigou, "Textural Analysis and Structure-Tracking for geological mapping: Applications to sonar data from Endeavour Segment, Juan de Fuca Ridge", Proc. OCEANS'93, IEEE-OES, p. 209-213, Victoria, B.C., 1993

Blondel, Ph.; "Segmentation of the Mid-Atlantic Ridge south of the Azores, based on acoustic classification of TOBI data", in MacLeod, C.J., P. Tyler and C.L. Walker (eds), "Tectonic, Magmatic and Biological Segmentation of Mid-Ocean Ridges", Geological Society Special Publication, no. 118, p. 17-28, 1996

Blondel, Ph., C. Sotin, Ph. Masson; "Adaptive Filtering and Structure-Tracking for statistical analysis of geological features in radar images", Computers and Geosciences, vol. 18, no. 9, p. 1169-1184, 1992

Cocquerez, J. P.; "Analyse d'images aériennes: extraction de primitives rectilignes et anti-parallèles", Ph.D. Thesis, Univ. Paris-Sud Orsay (France), 223 pp., 1984

Deriche, R.; "Using Canny's criteria to derive a recursively implemented optimal edge detector", Intern. Jour. of Computer Vision, vol. 1, p. 167-187, 1987

Duda, R.O., P.E. Hart; "Use of the Hough transformation to detect lines and curves in pictures", Comm. of A.C.M., vol. 15, p. 11-15, 1972

Duda, R.O., P.E. Hart; "Pattern recognition and scenes analysis", New York: Wiley, 1973

Fischler, M. A., J.M. Tenebaum, H.C. Wolf; "Detection of roadsand linear structures in low resolution aerial imagery using a multisource knowledge integration", Computer Graphics and Image Processing, vol. 15, p. 201-223, 1981.

Freeman, H.; "On the encoding of arbitrary geometric configurations", IEEE Trans. Elec. Computers, vol. EC-10, p. 260-268, 1961

Gonzalez, R.C., P. Wintz; "Digital Image Processing", Reading, MA: Addison-Wesley, 431 pp., 1977

Haralick, R.M., K. Shanmugam, R. Dinstein; "Textural features for image classification", IEEE Trans. Systems, Man, and Cybernetics, SMC-3, p. 610-621, 1973

Haralick, R.M.; "Statistical and structural approaches to texture", Proceedings of the IEEE, vol. 67, no. 5, 1979

Haralick, R.M., L.G. Shapiro; "Image Segmentation Techniques", Computer Vision, Graphics and Image Processing, vol. 29, p. 100-132, 1985

Hueckel, M.F.; "An operator which locates edges in digitized pictures", Jour. of A.C.M., vol. 18, p. 113-125, 1971

Jiang, M., W.K. Stewart, M. Marra; "Segmentation of Seafloor Sidescan Imagery using Markov random fields and neural networks", Proc. IEEE Oceans'93, vol. III, p. 456-461, 1993

Keeton, J.A.; "The use of image analysis techniques to characterise mid-ocean ridges from multibeam and sidescan sonar data", Ph.D. Thesis, 256 pp., Univ. Durham, UK, 1994

Kirsch, R. ; "Computer determination of the constituent structures of biological images", Comp. and Biomed. Res., vol. 4, p. 315-328, 1971

Lutz, T.M., J.T. Gutmann; "An improved method for determining and characterizing alignments of pointlike features and its implications for the Pinacate volcanic field, Sonora, Mexico", Journal of Geophysical Research vol. 100, no. B9, p. 17,659-17,670, 1995

Martelli, A. ; "An application of heuristic search methods to edge and contour detection", Comm. A. C. M. , vol. 19, no. 2, p. 73-83, 1976

Martelli, A., U. Montanari; "Optimal smoothing in picture processing; an application to fingerprints", Inf. Proc., vol. 71, p. 173-178, 1972

Meyer, F.; "Iterative image transformation for an automatic screening of cervical smears", Jour. Histochem. Cytochem., vol. 27, no. 1, p. 128-135, 1978

Michalopolou, Z.H., D. Alexandrou, C. de Moustier; "Application of neural and statistical classifiers to the problem of seafloor characterization", IEEE Journal of Oceanic Engineering, vol. 20, no. 3, p. 190-197, 1995

Mitchell, N.C., J.A. Spencer,; "An algorithm for finding routes for submarine cables or pipelines over complex bathymetry using graph theory", Hydrographic Journal, vol. 78, p. 29-32, 1995

Nevatia, R., K.R. Babu; "Linear features extraction and description", Computer Graphics and Image Processing, vol. 13, p. 257-269, 1980.

Nguyen, H.H., P. Cohen; "Gibbs random fields, fuzzy clustering, and the unsupervised segmentation of textured images", CVGIP: Graphical Models and Image Processing, vol. 55, no. 1, p. 1-19, 1993

Niblack, W.; "An introduction to digital image processing", Englewood Cliffs, NJ: Prentice Hall, 215 pp., 1986

Nilsson, N. J.; "Problem-solving methods in Artificial Intelligence", New York: McGraw-Hill, 1971

Paton, K.; "Line detection by local methods", Computer Graphics and Image Processing, vol. 9, p. 316-332, 1979

Prewitt, J.M.S.; "Object enhancement and extraction", in *Picture processing and psychopictorics*, B. S. Lipkins, A. Rosenfeld, New-York: Academic Press, p. 75-150, 1970

Reed, T.B., D. Hussong; "Digital image processing techniques for enhancement and classification of SeaMARC II side scan sonar imagery", J. Geophys. Res., vol. 94, no. B6, p. 7469-7490, 1989

Rhind, D.W.; "Global databases and GIS", in *The Association for Geographic Information Yearbook 1990*, M.J. Foster, P.J. Shand (eds), London: Taylor & Francis, p. 85-91, 1990

Rignot, E., R. Chellappa; "Segmentation of polarimetric Synthetic Aperture Radar data", IEEE Transactions on Image Processing, vol. 1, no. 3, 1992

Roberts, L.G.; "Machine perception of 3-dimensional solids", in *Optical and Electro-Optical Information Processing*, J.T. Tippett et al. (eds), p. 159-167 Cambridge, MA: M.I.T. Press, 1965

Robinson, G. S.; "Edge detection by compass gradient masks", Computer Graphics and Image Processing, vol. 6, p. 492-501, 1977

Rosenfeld, A., M. Thurston; "Edge and curve detection for visual scene analysis", IEEE Trans. Comp., vol. C-20, p. 562-569, 1971

Rosenfeld, A., M. Thurston, Y.H. Lee; "Edge and curve detection: further experiments", IEEE Trans. Comp., vol. C-20, p. 677-715, 1972

Sanfeliu, A., K.S. Fu; "A distance measure between attributed relational graphs for pattern recognition", IEEE Trans. on SMC, vol. 13, no. 3, p. 353-362, 1983.

Selim, S.Z., K. Alsultan; "A simulated annealing algorithm for the clustering problem", Pattern Recognition, vol. 24, no. 10, p. 1003-1008, 1991

Serra, J.; "Mathematical morphology", New York: Academic Press, 610 pp., 1982

Shapiro, S.D.; "Properties for transforms of curves in noisy pictures", Computer Graphics and Image Processing, vol. 8, p. 219-223, 1978

Shokr, M.; "Evaluation of second-order texture parameters for sea ice classification from radar images", Journal of Geophysical Research, vol. 96, no. C6, p. 10625-10640, 1991

Simmons, A.B., S.G. Chappell; "Artificial Intelligence - Definition and Practice", IEEE Journal of Oceanic Engineering, vol. 13, no. 2, p. 14-42, 1988

Sobel, I.; "Neighbourhood coding of binary images for fast contour following and general binary array processing", Computer Graphics and Image Processing, vol. 8, p. 127-135, 1978

Stewart, W.K., M. Jiang, M. Marra; "A neural network approach to sidescan imagery classification of a mid-ocean ridge area", IEEE Journal of Oceanic Engineering, vol. 19, no. 2, p. 214-224, 1994

Vanderbrug, G. J.; "Line detection in satellite imagery", IEEE Trans. Geoscience, vol. GE-14, p. 37-44, 1976

Wang, J. , P. J. Howarth; "Edge following as graph searching and Hough transform algorithms for lineament detection", Proc. IGARSS' 89, p. 93-96, Vancouver, 1989

Welch, T.A.; "A technique for high-performance data compression", IEEE Computer, vol. 17, no. 6, p. 8-19, 1984

Glossary of Abbreviations and Acronyms

AI	Artificial Intelligence
AR	Auto-Regressive model
ARMA	Auto-Regressive Moving Average model
AVG	Angle-Varying Gain
AVR	Axial Volcanic Ridge
AWI	Alfred-Wegener Institut, Germany
BP	Before Present
BP networks	Back-Propagation networks
BRIDGE	British Mid-Ocean Ridge Initiative
CAR	Circular Auto-Regressive model
DGPS	Differential Global Positioning System
DSL	Deep Submergence Laboratory, WHOI, USA
EPR	East Pacific Rise
EEZ	Exclusive Economic Zone
ETOPO5	Earth TOPOgraphy 5'-resolution
GIS	Geographic Information System
GLCM	Grey-Level Co-occurrence Matrices
GLORIA	Geological LOng-Range Inclined Asdic
GLORIA-B	GLORIA with Bathymetry
GPI	Geological and Paleontological Institute, Germany

GPS	Global Positioning System
HDF	Hierarchical Data Format
IFREMER	Institut Français pour l'EXploitation de la MER
ILSBL	Integrated Long and Short Baseline
JPL	Jet Propulsion Laboratory, USA
MAR	Mid-Atlantic Ridge
MARFLUX/ATJ	Mid-Atlantic Ridge Fluxes / Azores Triple Junction
MOR	Mid-Oceanic Ridge
MSG	Measurement-Space Guided clustering
NASA	National Air and Space Administration, USA
NGDC	National Geophysical Data Center, USA
RIDGE	Mid-Ocean Ridge Initiative, USA
RRS	Royal Research Ship
RV	Research Vessel
SBL	Short Baseline
SAR	Système Acoustique Remorqué (Towed Acoustic System)
SeaMARC	Sea Mapping And Remote Characterization
SIO	Scripps Institution of Oceanography, USA
SOC	Southampton Oceanography Centre, UK
SONAR	SOund Navigation And Ranging
TAG	Trans-Atlantic Geotraverse
TOBI	Towed Ocean-Bottom Instrument
TVG	Time-Varying Gain
USBL	Ultra-sHort Baseline
USGS	United States Geological Survey
UTM	Universal Transverse Mercator
UW	University of Washington, USA
WHIPS	Woods Hole Image Processing Software
WHOI	Woods Hole Oceanographic Institution
WWW	World-Wide Web

Index

abyssal plains and basins   3, 28, 48, 49, 129-152, 154, 155, 171, 190, 216, 238, 242, 267

accretionary prism   50, 53, 57, 58, 60, 62, 64, 65, 181

acoustic remote sensing
    sound propagation   5, 7, 8, 25, 28, 30, 60,
    interaction with the seafloor   7, 8, 32, 33

algae   179-181, 194, 195

anamorphosis   31, 42, 52, 53, 205, 227, 228, 237

anchor tracks   213, 214, 217

anthropogenic   194, 213, 216-221, 224, 243, 245, 256

Arctic   206, 207

artefact   3, 81, 93, 131, 162, 187, 220, 223-246

artificial intelligence   3, 248, 274, 275

avalanche   *see* slide

AVG   29, 43, 226, 236

banding   140-142, 151

bathymetry   25, 31, 42, 43, 75, 77, 85-87, 104, 118, 123, 125, 131, 132, 178, 188, 235, 248, 265, 266, 268, 271

backscattering models   32, 33, 43, 45, 46, 78, 97, 265

Baltic Sea   194, 195, 197, 205, 207, 208, 212-214, 219

beam spreading   233, 234

bioherms   179-180

biological activity   179-181, 194, 210-212, 225, 242, 268, 271

"black smokers"   116-120

Black Sea   181, 183, 185, 245

blocks   77, 109, 114-116, 136, 137, 144, 145, 148-150

bottom currents   131, 137, 157, 167, 180, 194-196, 200, 201, 204, 205, 209, 212,
        217, 224

brine structures   186-188

cable   20, 129, 151, 154, 186, 211, 245, 269, 275

Canaries   137

canyon   48, 133, 135, 146, 148, 149, 155, 157-163, 165-167, 173-176, 237

carbonate   81, 155, 156, 172, 173, 179, 203, 204

channel   54-56, 59, 60, 65, 133, 138, 139, 148-151, 155, 157-159, 163, 169, 170,
        173, 174, 202, 204, 241, 256

Chile   167, 174, 190

clay   199, 200

coastal environments   3, 60, 61, 145, 168, 179, 181, 190, 191, 193-221, 243

collision margins   47, 62-64, 69

Columbian Trench   54-56, 59-61, 65-67

continental margins   59, 130, 133, 148, 153-191, 194, 207, 237, 243

    shelf   3, 48, 49, 154, 157, 159-161, 174, 179-181, 196, 242

    slope   3, 133, 134, 154, 155, 157-159, 161, 162, 174

    rise   48, 49, 154, 155, 157, 161, 163, 169, 170

contour analysis   256-262

contourite   131, 137, 155, 168

contrast enhancement   *see* histogram manipulation

coral   156, 179, 194, 210-212

cross-talk   10, 43, 220, 225

debris   131-133, 137-140, 143, 147-151, 157, 163, 165, 178, 183, 185

deformation front   53-55, 57, 58

Doppler effect   8, 18

DSL-120 sonar   12, 20, 42, 43, 44, 75-77, 117-120, 224, 225, 238, 256, 258, 266

drop-out lines   29, 42, 43,

dump site   219

East Pacific Rise   100, 101, 112,113, 119, 269, 270

edge detection   257-260

EG&G 272T sonar   12

EG&G990S sonar   184, 185, 190

EG&G Deep-Tow sonar   12

erosion   48, 63, 137, 143, 155, 159, 172-174, 199, 204, 243

evaporite   186, 188

expert systems   3, 274, 275

fan   48, 59, 60, 133-135, 155, 162

fish   212, 213, 225, 242

fjord   197, 206, 207, 220

flows   154, 157, 163, 182-185, 206

footprint   10, 11, 249

Fourier transform   255, 256

geo-referencing   34

GIS   42, 248, 267-271

glacial   194, 195, 206, 207, 220

GLORIA sonar   12, 21, 41, 43, 54-57, 60-67, 69, 104-106, 120-126, 133-135, 140,
        146-148, 150, 155-161, 163, 164, 168-173, 177, 178, 182, 190, 196, 224-226, 234,
        238-240, 242, 256, 271

GLORIA-B sonar   12, 104

GPS   16, 17, 20, 36, 42, 93,

graben   84, 119, 120

gradient filtering   255-260

gravel   196, 200, 204

grazing angle   7

Hawaii   120, 135, 146-148, 150, 238

harbour   194, 213, 214, 219, 269

heave   24, 228, 229, 236

histogram
    manipulation   23, 38, 39, 251-253
    statistics   36, 37, 251, 252

HMR-1 sonar   *see* SeaMARC II

hummocky ridge   78, 96-101, 104, 105, 108,

hummocky terrain   79, 80, 82, 83, 87, 89-97, 99, 100, 102, 104, 108, 114-116, 249,
    254, 262, 263, 272

hydrothermalism   51, 74, 116-120, 265, 267, 269-271

  deposits   74, 117-120, 267, 269, 270

  mounds   75-77, 117-120

  vents   74, 116-120, 224, 256, 266, 270, 271

iceberg   194, 206, 225

image processing 3, 247-276

incidence (angle of)   7, 86

Indian Ocean   123-126

Indonesian Margin   179-181

interference   188, 224, 225, 238-240, 245

interpolation   10, 27, 35, 42, 46

interpretation (checklist)   51-53

IZANAGI sonar   57, 58, 59

Jason 200 kHz sonar   12

Juan de Fuca   119, 120

Kermadec Trench   123

Klein 520 sonar   12

Klein 590/595 sonar   12

lake   194, 209, 210

Lambert law   33

Lau   120, 121, 123

lava flows   74, 79, 83, 84, 89, 91, 94, 95, 105, 111, 112, 120, 123, 125, 148-150, 238,
    257, 264, 266, 270

layback   18, 25

layover   88, 234, 235, 245

levee   163, 166, 206

Levitus database   6, 28

Lloyd's effect   239, 240

lobate flows   82, 83, 89, 96

Madeira abyssal plain   135, 138, 139, 144, 145, 240

Magellan spacecraft   1, 33, 244

MAK-1 sonar   12, 170, 171, 182, 183, 186, 189, 196, 245

Marianas   49, 67, 68, 69

Mars   145, 146, 223, 243, 244

mass-movement   130, 151, 155, 160, 163, 165, 174, 190

meander   162, 163, 174, 241

Mediterranean Sea   170, 171, 181, 184-187, 190, 245

megaripple   202-204

Mexico (Gulf of)   133-135, 154

Mid-Atlantic Ridge   74, 77, 82-84, 88, 94-97, 100, 104-106, 108, 109, 112-115,
          117-119, 131, 177, 244, 265, 271, 272

mid-ocean ridge   3, 28, 48, 49, 73-126, 130, 131, 148, 153, 154, 170, 177, 224, 229,
          242, 244, 265, 272

Mississippi   133-135

mosaicking   3, 23, 34, 35, 42-44, 99, 100, 102, 106, 134, 160, 163, 171, 253, 272

mud   81, 107, 132, 140, 156, 172, 182-186, 188, 195-197, 201, 207, 209, 213-216,
          219, 220, 245

mud vent   182, 183

mud volcanism   50, 65-67, 181-188, 245

multiple   40, 41, 240-242

Nankai Trough   57-59, 181

navigation
    ship   16, 17, 24
    towfish   18, 19, 20, 25

noise   11, 23, 26, 39, 42, 43, 65, 78, 83, 96, 188, 243, 253, 255, 256, 272

North Sea   154, 155, 190, 194, 205, 217

offsets
    transform   79, 107, 113
    non-transform   79, 113, 115, 116, 273

OKEAN sonar   12

otter-board   197, 214, 216, 220

outcrop   156, 173, 187, 197-199, 211, 220, 256-258

penetration depth   131, 132, 148, 238

Peru   163, 167, 174, 190, 242

Peru-Chile Trench   59, 61-63, 174

pillow lavas   78-80, 89, 94, 96, 112,

pipeline   20, 129, 151, 186, 245, 262, 268, 269, 275

pitch   20, 26, 52, 53, 231-233, 236

ploughmark (iceberg)   194, 206

pockmark   171, 188-190, 207-210,

positioning   18, 19, 20, 21, 24, 25, 26

plate tectonics   48

pressure ridge   140, 141

Puerto Rico Trench   64, 65

quantisation   27, 41, 236, 250

radar   5, 33, 46, 238, 244, 248

relict structures   144-146, 151, 194

roll   20, 26

reflection (angle)   7

resolution   10, 11

Reykjanes Ridge   9, 79, 80, 81, 91, 92, 102, 103, 105, 107

ripple   194, 195, 199-204, 209, 216, 221

roll   52, 53, 230-233, 236

roughness (micro-scale)   8, 32, 33, 34, 87, 94, 96, 156,

rubbersheeting   35, 44

salinity   6, 224

sand   195, 196, 199-205, 212, 215-220

sand ribbons   200, 204

SAR sonar   12

scour   156, 173, 178, 198, 205, 206

scree   80, 83-85, 87, 88, 96, 106, 114-116, 138,

seafloor characterisation   262-265, 271-273, 275

seagrass   see vegetation

SeaMARC I sonar   99-101, 106, 112, 113, 165

SeaMARC II sonar   12, 68, 69, 79-81, 90-93, 107, 162, 163, 166-168, 174-176, 190, 239, 241, 242

seamount   62, 63, 131, 144, 148-151, 177, 178, 189, 235, 242, 244, 249, 251, 256, 266

sediment   34, 41, 54, 59, 63-65, 80, 81, 83, 84, 87-93, 95, 96, 98-105, 107, 111, 112,
        114-116, 123, 130, 131, 133, 136, 138, 140, 143, 144, 148-150, 154-157, 159, 161,
        163, 166, 167, 169, 171, 172, 177-179, 182, 187, 188, 191, 194, 195, 199-202,
        204, 205, 207, 209-217, 238-240, 242, 252, 257, 263, 269, 272, 273

sediment waves   167-169, 174, 200, 201

seepage   181, 188, 224

serpentinite   51, 67, 68, 106, 114-116, 267

serpentinite seamounts   51, 65, 67-69

sheet flows   78-80, 83, 85, 89, 90, 101, 102, 104, 112

shipwreck   20, 151, 216-218

Simrad MS-992 sonar   12

Simrad EM-12 sonar   12, 20

slant-range   7, 8, 30, 31, 41-43, 52, 53, 88, 92, 93, 95, 169, 204, 234, 235, 237, 241,
        265

slide   134-137, 140, 147, 148, 151, 155, 160, 161, 163-166, 173, 183, 206, 257, 269

slump   50, 58, 59, 62, 138, 139, 163, 169, 170, 172, 174, 179, 206, 207

SMS-960 sonar   179-181

sound velocity   5, 6, 10, 28, 238

speckle   *see* noise

splay   106, 108, 109, 111-113, 163,

SPOT satellite   29

stencilling   34, 35, 99, 100, 102, 208, 253

striping
        across-track   29, 231, 236, 238
        along-track   29, 66, 225, 236, 238

subduction   47, 49-51, 54, 63, 67, 123, 154, 174, 177, 190

surface reflection   230, 241

surface waves   194-196, 200, 201, 212, 215, 218

TAG mound   75-77, 117-120

talus   82. 87, 89, 98, 115, 272

[TAMU]^2 sonar   12

tectonic structures   54-59, 62, 63, 65, 66
        fault   50, 74, 79, 84, 93, 103-112, 114, 125, 160, 188, 249, 263, 266

fissure   74, 79, 83-85, 95, 96, 104, 106, 107, 109, 111, 112, 120, 256-258, 263, 269, 270

scarp   54-56, 74, 77-79, 83, 95, 96, 99, 101, 102, 106-113, 115, 116, 123, 124, 144, 148, 149, 165, 166, 171, 188, 237, 249-252, 254, 264, 265, 269, 272

temperature   6, 224

terrace   98, 163, 166, 167, 174, 251

texture analysis   248, 256, 262-265, 273

thresholding   108, 110, 112, 125

till   194, 195, 204

TOBI sonar   9, 12, 20, 21, 25, 35, 36, 41, 43, 44, 77, 79-85, 87-90, 92-100, 102, 105, 106, 108, 111, 112, 114, 115, 117, 118, 134, 135, 138-145, 170, 187, 188, 190, 223-225, 228-230, 239, 240, 244, 248-250, 254, 257, 258, 263, 265, 271-273

trawl marks   151, 214, 215, 243

trench   3, 47-50, 131

tsunami   50

turbidite   48, 131-134, 137-139, 151, 155, 157, 162, 163, 170, 173, 174

TVG   28, 29, 41-43, 52, 53, 188, 204, 225, 226, 236

vegetation   179, 180, 194

Venus   1, 33, 244

volcano

AVR   79, 102-106, 229, 244

chain   74,

cluster   79, 90-93, 95

composite   82, 93, 94

flat-topped   83, 87-92, 94, 95, 100, 101, 103, 105, 249, 272

point-source   78-80, 83-85, 90, 94, 96

hummocky ridge   78, 96-101, 104, 105, 108,

summit crater   79-85, 87, 89, 90, 93, 94, 98, 100, 101, 105, 112, 178, 181, 182

volume reverberation   32, 131, 226, 228, 238

water column   199, 202, 203, 217, 224, 225, 229, 271

waves

sediment   *see* sediment waves

surface   *see* surface waves

yaw   20, 52, 53, 93, 231-233, 236